Synthesis Methods and Crystallization

Edited by Riadh Marzouki

Published in London, United Kingdom

IntechOpen

Supporting open minds since 2005

Synthesis Methods and Crystallization
http://dx.doi.org/10.5772/intechopen.77471
Edited by Riadh Marzouki

Contributors
Riadh Marzouki, Youssef Ben Smida, Savas Kaya, Sultan Erkan, Mohamed Faouzi Zid, Ahmed Hichem Hamzaoui, Francisco Rey-García, Carmen Bao-Varela, Florinda Costa, Mei Luo, Shafirah Samsuri, Wan Nur Athirah Mazli, Mohd Afnan Ahmad, Veronika Rozhdestvina, Xinwen Guo, Yitong Han, Hong Yang

First published in London, United Kingdom, 2020 by IntechOpen
IntechOpen is the global imprint of INTECHOPEN LIMITED, registered in England and Wales, registration number: 11086078, 5 Princes Gate Court, London, SW7 2QJ, United Kingdom
Printed in Croatia

British Library Cataloguing-in-Publication Data
A catalogue record for this book is available from the British Library

Additional hard and PDF copies can be obtained from orders@intechopen.com

Synthesis Methods and Crystallization
Edited by Riadh Marzouki
p. cm.
Print ISBN 978-1-83880-223-3
Online ISBN 978-1-83880-224-0
eBook (PDF) ISBN 978-1-83968-901-7

For EU product safety concerns:
IN TECH d.o.o., Prolaz Marije Krucifikse Kozulić 3, 51000 Rijeka, Croatia, info@intechopen.com or visit our website at intechopen.com.

Meet the editor

Riadh Marzouki is Assistant Professor in International Cooperation between the Department of Chemistry, Faculty of Sciences at King Khalid University (Saudi Arabia); assistant professor in the Department of Chemistry, Faculty of Sciences, University of Sfax (Tunisia); and researcher in the Laboratory of Materials and Crystallography of the University of Tunis El-Manar (Tunisia). His research activities include investigating the synthesis, crystal structure, electrical properties, and biological activities of new crystalline materials. He is also the author of thirty-six papers and three book chapters.

Contents

Preface XI

Section 1
Introduction 1

Chapter 1 3
Introductory Chapter: Crystalline Materials and Applications
by Riadh Marzouki

Section 2
Synthesis Methods 7

Chapter 2 9
Synthesis Methods in Solid-State Chemistry
by Youssef Ben Smida, Riadh Marzouki, Savaş Kaya, Sultan Erkan, Mohamed Faouzi Zid and Ahmed Hichem Hamzaoui

Chapter 3 25
One-Pot Synthesis of Chiral Organometallic Complexes
by Mei Luo

Section 3
Nucleation and Crystal Growth 47

Chapter 4 49
Nucleation
by Wan Nur Athirah Mazli, Mohd Afnan Ahmad and Shafirah Samsuri

Chapter 5 59
Synthesis Methods and Crystallization of MOFs
by Yitong Han, Hong Yang and Xinwen Guo

Chapter 6 81
Laser Floating Zone: General Overview Focusing on the Oxyorthosilicates Growth
by Francisco Rey-García, Carmen Bao-Varela and Florinda M. Costa

Chapter 7 103
Solid Phase Evolution of Nanodispersed Palladium Powders
by Veronika Ivanovna Rozhdestvina

Preface

Considering the diversity of the crystalline structures of crystallized materials and their impact on these materials' physical properties, this book describes the synthesis methods most adopted for the preparation of crystalline phases. Many types of materials have been processed, including inorganic, organic, organometallic, and hybrid materials. Microstructure is one parameter influencing physical properties of crystals while nucleation and crystallization are others. In fact, crystal growth influences the size and morphology of crystals. Therefore, several examples of materials have been controlled by scanning or transmission electron microscopy as a function of the parameters influencing the microstructure, particularly, time, temperature, reaction medium, and so on.

This book is the result of several years of scientific research in crystallography and solid-state chemistry. We thank all who have contributed to this work from near and far, especially Professors Mohamed Faouzi Zid and Ahmed Driss.

Riadh Marzouki
Chemistry Department,
College of Science,
King Khalid University,
Abha, Saudi Arabia

Chemistry Department,
Faculty of Sciences of Sfax,
University of Sfax,
Tunisia

Laboratory of Materials,
Crystal Chemistry and Applied Thermodynamics,
LR15ES01,
Faculty of Sciences of Tunis,
University of Tunis El Manar,
Tunisia

Section 1

Introduction

Chapter 1

Introductory Chapter: Crystalline Materials and Applications

Riadh Marzouki

1. Introduction

The scientific research and development in crystalline materials science covers the synthesis, crystal structure study, physicochemical properties, and applications of solid crystals (inorganic, organic, hybrid, and organometallic). Indeed, the synthesis and the physicochemical characterization of solid materials with interesting physical properties can bring us back to the specific applications related to their crystalline structures. Accordingly, technological development necessitates the exploration of new crystalline materials.

Currently, energy storage, water treatment, and the synthesis of pharmaceutical products are among the areas of intense research activity in materials chemistry.

Given the high energy demand which continues to increase, with a progressive depletion of fossil fuels, which lead to the emission of greenhouse gases and their environmental impact (greenhouse effect), current scientific research is focused on renewable energy research areas and also energy storage. In this context, the investigation of rechargeable batteries constitutes a system of storage of electrical energy in chemical energy. In fact, the research of new crystalline materials with open frameworks formed by diverse polyhedra (tetrahedra, pentahedra, and octahedra) bounding interlayer spaces and/or tunnels communicating through the intermediary of windows where cations are located is currently a field of intense activity in energy storage, especially Li-ion batteries and Na-ion batteries, etc. [1–9]. The preparation and characterization of new compounds are the driving forces of recent technological development, and studies are progressing through the exchange of views between relevant specialists. On the other hand, new research works on the treatment of batteries are based on degradable polymers such as alkali-cellulose in order to avoid materials based on expensive and less exploited metals (Li) [10].

2. Crystalline materials and applications

As part of the crystalline materials research for biologically active molecules, groups of organic heterocycles appear, such as coumarin derivatives [11, 12], which can be found in the plant kingdom. The synthesis and structural characterization work carried out on these materials constitutes a vast field of research, in which a large number of laboratories are involved in the world. Natural or synthetic compounds are particularly sought after for their biological and pharmaceutical activities. These compounds have been used in different fields of application such as food additives and dyes, as well as in the cosmetics field. In fact, these phases are powerful antioxidants and have antibacterial, hypolipidemic, cholesterol-lowering, and anticarcinogenic properties, which give them a great importance in the pharmaceutical field. Charge transfer agents, solar energy collectors and non-linear optical materials require materials with inherent photochemical characteristics, reasonable stability and good solubility in some.

Given the enormous increase in industrial activity, enormous quantities of polluting chemical wastes are spreading in the air, soil, and air [13]. In fact, every year more than 2 million gallons of synthetic dyes of which more than 10% are lost in waterways during their use [14]. However, the presence of organic compounds, unwanted metals, and dye waste requires treatments before use. In this context, the photocatalytic treatment of water is one of the treatments used due to the great precision of mineralization of organic compounds, and the ease of the process, in addition to the advantages of replacing filters with all the problems. Regeneration and fouling are associated with their use. Several crystalline materials have been introduced into the filters of the catalytic treatments [15, 16]. These materials have been used to eliminate various types of pollutants such as organic compounds, heavy metals, and others.

This book is dedicated to the investigation of various types of crystalline materials that can be used in several fields nowadays, such as energy storage (rechargeable batteries). Other crystalline compounds can be used in photocatalytic treatments. Other pharmaceutical compounds are also processed due to their interesting biological activities. The synthesis method of each compound is detailed to facilitate those interested in its preparation. As the physical and biological properties of the compound are related to the crystal structure, a description of the structure of each compound has been discussed.

Author details

Riadh Marzouki[1,2,3]

1 Chemistry Department, College of Science, King Khalid University, Abha, Saudi Arabia

2 Laboratory of Materials, Crystal Chemistry and Applied Thermodynamics, LR15ES01, Faculty of Sciences of Tunis, University of Tunis El Manar, Tunisia

3 Chemistry Department, Faculty of Sciences of Sfax, University of Sfax, Tunisia

*Address all correspondence to: rmarzouki@kku.edu.sa

References

[1] SRS P, Michael MS, Radhakrishna S, Julien C. Novel low-temperature synthesis and characterization of $LiNiVO_4$ for high-voltage Li ion batteries. Journal of Materials Chemistry. 1997;7:1791

[2] Marzouki R, Smida YB, Sonni M, Avdeev M, Zid MF. Synthesis, structure, electrical properties and Na^+ migration pathways of $Na_2CoP_{1.5}As_{0.5}O_7$. Journal of Solid State Chemistry. 2020;**285**:121058

[3] ALQarni OSA, Marzouki R, Ben Smida Y, Alghamdi MM, Avdeev M, Belhadj Tahar R, et al. Synthesis, structure, electrical properties and Na^+ migration pathways of $Na_2CoP_{1.5}As_{0.5}O_7$. PRO. 2020;**8**(3):305

[4] Marzouki R. Electrical properties and alkali-pathways simulation of new mixed conductor $Na_4Li_{0.62}Co_{5.67}Al_{0.71}(AsO_4)_6$. Materials Research Express. 2020;7:016313

[5] Marzouki R, Zid MF. Non-centrosymmetric $Na_7Li_{0.8}K_{0.2}CO_5(As_3O_{10})_2(As_2O_7)_2$: Synthesis, structure and alkali ion-conduction pathways simulation. International Journal of Electrochemical Science. 2020;**15**: 3776-3792

[6] Moussa MAB, Marzouki R, Brahmia A, Georges S, Obbade S, Zid MF. Synthesis and structure of new mixed silver cobalt(II)/(III) diphosphate - $Ag_{3.68}Co_2(P_2O_7)_2$. Silver(I) transport in the crystal. International Journal of Electrochemical Science. 2019;**14**(20): 1500-1515

[7] Nasri R, Marzouki R, Georges S, Obbade S, Zid MF. Synthesis, sintering, electrical properties and sodium migration pathways of new lyonsite $Na_2Co_2(MoO_4)_3$. Turkish Journal of Chemistry. 2018;**42**(5):1251-1264

[8] Marzouki R, Smida YB, Guesmi A, Georges S, Ali IH, Adams S, Zid MF. Structural and electrical investigation of new melilite compound $K_{0.86}Na_{1.14}CoP_2O_7$". International Journal of Electrochemical Science. 2019;**13**:11648-11662

[9] Marzouki R, Sayed MA, Graia M, Zid MF. Cobalt Compounds and Applications. London, UK: IntechOpen; 2019. DOI: 10.5772/intechopen.86215

[10] Marzouki R, Brahmia A, Bondock S, Keshk SMAS, Zid MF, Al-Sehemi AG, et al. Mercerization effect on structure and electrical properties of cellulose: Development of a novel fast Na-ionic conductor. Carbohydrate Polymers. 2019;**221**:29-36

[11] Bejaoui L, Brahmia A, Marzouki R, Dusek M, Eigner V, Serdaroğlu G, et al. Synthesis, crystal structure, hirshfeld surface analysis, spectroscopic, biological and first-principles studies of novel aminocoumarins. Journal of Molecular Structure. 2020;**1221**:128862. DOI: 10.1016/j.molstruc.2020.128862

[12] Brahmia A, Marzouki R, Rohlicek J, Irfan A, Al-Sehemi AG, Hassen RB. Structural, spectroscopic and first-principles studies of new amino coumarin derivatives. Acta Crystallographica Section C: Structural Chemistry. 2019;**75**:1617-1627

[13] Gavrilescu M, Demnerová K, Aamand J, Agathos S, Fava F. Emerging pollutants in the environment: Present and future challenges in biomonitoring, ecological risks and bioremediation. New Biotechnology. 2015;**32**:147

[14] Houas A, Lachheb H, Ksibi M, Elaloui E, Guillard C, Herrmann J-M. Photocatalytic degradation pathway of methylene blue in water. Applied Catalysis. 2001;**B31**:145

[15] Marzouki R, Abd-Rabboh HSM, Baker AH, Ghazwani SA, Zid MF, Hamdy MS. Synthesis, characterisation and the photocatalytic performance of europium oxide/ceria nanocomposite. International Journal of Environmental Analytical Chemistry. 2019. DOI: 10.1080/03067319.2019.1694671

[16] Abd-Rabboh HSM, Marzouki R, Alassaf A, Loghbi M, Hamdy MS. Removal of malachite green dye from contaminated aqueous solutions using WO_3/Eu_2O_3-visible-light-assisted photocatalysis. International Journal of Environmental Analytical Chemistry. 2019. DOI: 10.1080/03067319.2019.1683551

Section 2

Synthesis Methods

Chapter 2

Synthesis Methods in Solid-State Chemistry

Youssef Ben Smida, Riadh Marzouki, Savaş Kaya, Sultan Erkan, Mohamed Faouzi Zid and Ahmed Hichem Hamzaoui

Abstract

The synthesis of single crystal is an area of intense activity in the materials science. The obtaining of the single crystal with sufficient dimension for X-ray diffraction depends on several factors including the chemical composition, crystal structure of the reagents, and physical parameters (temperature and pressure). In this context, this chapter is dedicated to the description of the most common synthesis methods of single crystal in the solid-state chemistry: solid-state method, hydrothermal, and slow evaporation at room temperature. Same other materials can be obtained at high pressure. There are also some physical techniques to grow single crystal, each technique is specific for specific materials.

Keywords: synthesis, single crystal, hydrothermal, evaporation, crystal growth

1. Introduction

Synthesis is the most important step in solid-state chemistry research and in materials science. The samples may be prepared as single crystals, polycrystalline powder, or a thin film. Glass and amorphous samples are from another class of materials. In this chapter, we discuss only the methods of synthesis of single crystals.

The discovery of new materials passes necessarily by the X-ray single crystal method. Usually, the polycrystalline powder and thin film are designed for the known materials, where their structures have been determined by X-ray single crystal diffraction. It is possible to determine the crystal structure of the polycrystalline samples, but generally it shows some difficulty. However, powder XRD is generally dedicated to materials that cannot be prepared as single crystals like some pharmaceutical compounds.

The four most adopted methods of synthesis of single crystals are solid-state, hydrothermal, slow evaporation at room temperature, and flux methods. Here, we discuss also the crystal structure prediction method at high pressure. Each method is controlled by several controllable and/or uncontrollable parameters. The parameters influence the obtaining of phase, morphology, and size of single crystals. The most important criterion in determining the crystal structure is the size of the single crystal and the crystallinity. These two parameters may be checked before the X-rays single crystal diffraction by using the binocular magnifying glass and the polarizing microscope.

2. Solid-state method

The solid-state method is the simplest method. The number of acting parameters is relatively few but difficult to control. This method is commonly used for the synthesis of the single crystals and polycrystalline powders of phosphates and arsenates of transition metals and monovalent cations [1–17].

The crystallization is a phenomenon which generally occurs during a phase change and accompanied by a thermal effect. It is carried out in two stages: germination and growth. The germination begins at a point where the phases are not in equilibrium, a condition can be favored by several factors such as crucible wall, impurity, amorphous. This step consists of the appearance within the reaction mixture of "germs." The growth takes place in several stages: reorganization of the atoms, adsorption on the surface of the solid, diffusion, and fixation of the atoms on their final sites. Successive layers therefore aggregate on the faces of the crystal which sees its volume increase.

The solid-state synthesis can be done into two steps:

Preliminary treatment: This step consists in weighing the desired quantities of the precursors and then grinding them in an agate mortar. The powder obtained is placed in a crucible or a porcelain basket (**Figure 1**) and preheated between 350 and 400°C for a few hours (**Figure 1**) This operation allows the decomposition of the starting reagents and removes the volatile products such as NH_3, NO_2, CO_2, and

Figure 1.
Equipment used in solid-state synthesis.

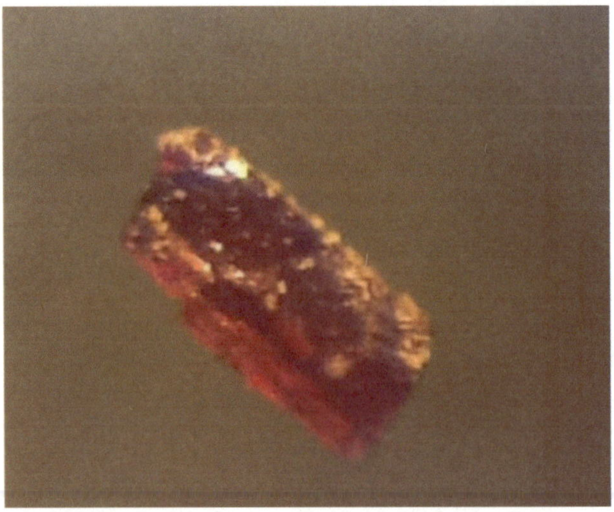

Figure 2.
Image of a single crystal of the phase $Na_{1.25}Co_{2.187}Al_{1.125}(AsO_4)_3$ [9] seen under the binocular magnifying glass.

Single crystal	Reagents	T1 (°C)	T2 (°C)	t1	t2	R	Ref.
$LiCo_2As_3O_{10}$	$Li_2CO_3 + CoCl_2.6H_2O + NH_4H_2AsO_4$	350	730	12 h	3days	5 K/h	[2]
$NaCo_2As_3O_{10}$	$NaNO_3 + Co(NO_3)_2 .6H_2O + As_2O_5$	400	670	24 h	3 days	5 K/h	[3]
$Na_4Co_7(AsO_4)_6$	$NaNO_3 + Co(NO_3)_2 .6H_2O + NH_4H_2AsO_4$	350	750	24 h	3 days	5 K/h	[4]
$K_{0.13}Na_{3.87}MgMo_3O_{12}$	$Na_2CO_3 + K_2CO_3 + (NH_4)_2Mo_4O_{13} + Mg(NO_3)_2.6H_2O$	400	600	12 h	5 days	5 K/h	[5]
$Na_{2.77}K_{1.52}Fe_{2.57}(AsO_4)_4$	$Na_2CO_3 + K_2CO_3 + Fe(NO_3)_3.9H_2O + NH_4H_2AsO_4$	400	850	24 h	3 days	5 K/h	[6]
$K_{0.86}Na_{1.14}CoP_2O_7$	$NaNO_3 + KNO_3 + Co(NO_3)_2.6H_2O + NH_4H_2PO_4$	400	660	12 h	4 days	5 K/h	[7]
$Na_2CoP_{1.5}As_{0.5}O_7$	$NaNO_3 + Co(NO_3)_2.6H_2O + NH_4H_2PO_4$	350	620	12 h	3 days	5 K/h	[8]
$Na_{1.25}Co_{2.187}Al_{1.125}(AsO_4)_3$	$Na_2CO_3 + Co(NO_3)_2 .6H_2O + NH_4H_2AsO_4$	350	800	12 h	3 days	5 K/h	[9]
$Na_3Co_2(As_{0.52}P_{0.48})O_4As_2O_7$	$Na_2CO_3 + Co(NO_3)_2.6H_2O + NH_4H_2AsO_4 + NH_4H_2PO_4$	350	700	12 h	3 days	5 K/h	[10]
$K_{0.405}Bi_{0.865}AsO_4$	$K_2CO_3 + Bi_2O_3 + NH_4H_2AsO_4$	400	850	12 h	30 days	5 K/h	[11]
$(Na_{0.71}Ag_{0.29})_2CoP_2O_7$	$NaNO_3 + AgNO_3 + Co(NO_3)_2{\cdot}6H_2O + NH_4H_2PO_4$	400	620	24 h	5 days	5 K/h	[12]
$Ag_4Co_7(AsO_4)_6$	$AgNO_3 + Co(NO_3)_2{\cdot}6H_2O + As_2O_5$	400	1005	24 h	5 days	5 K/h	[13]
$Na_4Li_{0.62}Co_{5.67}Al_{0.71}(AsO_4)_6$	$Na_2CO_3 + LiOH.H_2O + Co(NO_3)_2.6H_2O + Al_2O_3 + NH_4H_2AsO_4$	400	870	24 h	5 days	5 K/h	[14]
$Ag_{3.68}Co_2(P_2O_7)_2$	$AgNO_3 + Co(NO_3)_2{\cdot}6H_2O + NH_4H_2PO_4$	350	610	12 h	3 days	5 K/h	[15]
$Na_2Co_2(MoO_4)_3$	$NaNO_3 + Co(NO_3)_2{\cdot}6H_2O + (NH_4)_6Mo_7O_{24}$	350	650	24 h	4 days	5 K/h	[16]
$Na_7Li_{0.8}K_{0.2}Co_5(As_3O_{10})_2(As_2O_7)_2$	$NaNO_3 + Co(CH_3COO)_2.4H_2O + As_2O_5$	400	670	12 h	7 days	5 K/h	[17]

T1, pre-treatment temperature; T2, temperature of synthesis; t1, pre-treatment time; t2, synthesis time; R, cooling rate.
The different parameters of the synthesis (reagents, pre-treatment temperature, temperature of synthesis, pre-treatment time, synthesis time, and cooling rate) are regrouped in the table.

Table 1.
Same materials have been obtained as single crystals by the means of solid-state reaction.

H_2O, only the oxides remain. The mixture is again ground at the outlet of the oven to make it more homogeneous and to minimize the grain size.

Crystal growth: After the germination phase, and under the effect of a concentration gradient, the cations have just migrated to the germs, forming well-ordered layers. This migration is favored by heating at very high temperature. After cooling, the crystals are separated from the stream by hot and sometimes boiling water.

The disadvantages of this method are that it is very slow and needs a lot of energy. In fact, the reaction occurs at high temperatures (500–2000°C) for several hours and for same time for several days. The heating at these temperatures may decompose the desired compound.

Experimentally, oxides and nitrates are bad reagents in the synthesis of single crystals, and they often give crystals with small size which is insufficient to do the x-ray single crystal diffraction. The mechanical grinding can be used to decrease the grain sizes and increase the specific surface then increase the reactivity.

The cooling rate is a very important factor to obtain a single crystal with good crystallinity. The cooling rate should be as slow as possible and at least up to 50°C below the crystallization temperature. The choice of the size and the confirmation of the crystallinity of single crystals are initially done using a binocular magnifier (**Figure 2**) then by using the polarizing microscope. This choice is confirmed by the intensity and the width of the diffracted X-rays.

Table 1 summarized that same materials have been obtained as single crystals by the means of solid-state reaction. The different parameters of the synthesis (reagents, pre-treatment temperature, temperature of synthesis, pre-treatment time, synthesis time, and cooling rate) are regrouped in the table.

The resolution of the structure same crystals needs the knowledge of its compositions by using elementary analysis such as the energy-dispersive X-ray spectroscopy (EDX) (**Figure 3**).

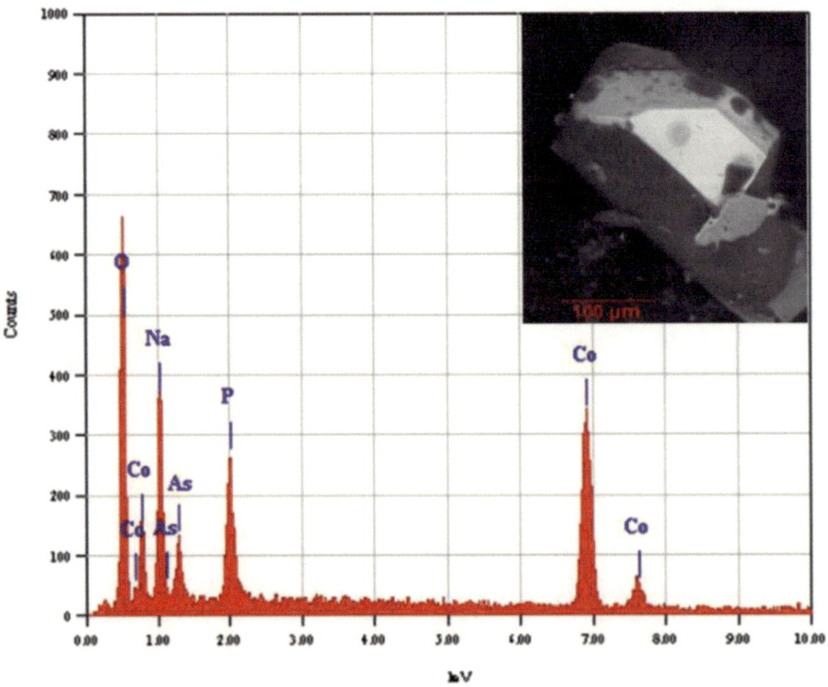

Figure 3.
SEM micrograph and EDX analysis of a single crystal of $Na_2CoP_{1.5}As_{0.5}O_7$ [8] showing the morphology, size, and composition of the single crystal.

3. Hydrothermal method

The synthesis of the single crystal by the means of hydrothermal method occurs usually in water at temperatures between 180 and 300°C. The reactor can be an autoclave (**Figure 4**) or a sealed glass tube (**Figure 5**). The pressure is controlled by the gas law [P = *f*(T)]. The pressure of same reactors can be controlled, and it can reach a value of 850 GP. Several materials have been synthesized using the hydrothermal method.

The hydrothermal conditions of an aqueous medium correspond to temperatures and pressures above 100°C and 1 bar, respectively. These conditions allow to considerably modify the chemistry of the cations in solution. They favor the formation of complex metastable structures of lower symmetry and involving smaller variations in enthalpy and entropy than under "normal" conditions [18, 19]. Hydrothermal conditions are also those of the geological processes during which many minerals were formed. In the laboratory, such conditions are achieved by heating a solution in a closed enclosure (autoclave and sealed glass tube) at temperatures of the order of 200–400°C.

The thermodynamic properties of water up to temperatures of 1000°C and pressures of several tens of kilobars are well known [18]. Quantitative data are collected in numerous review articles [18–21]. There are three essential points to remember.

- The dielectric constant of water drops when the temperature increases. It increases by pressure increase [22] (**Figure 6**). The hydrothermal solutions are therefore characterized by low dielectric constants and the electrolytes which are completely dissociated under normal conditions preferentially form pairs of ions or complexes of low electrostatic charge.

- The viscosity of water decreases with the increase of temperature [23], which leads to greater mobility of the dissolved species than under normal conditions.

- The ionic product of water increases strongly with temperature [24] (**Figure 7**). The conductivity measurements allow establishing the law of variation of the ionic product as a function of temperature

$$\log K_e = -(3018 / T) - 3.55 \qquad (1)$$

Figure 4.
Autoclave.

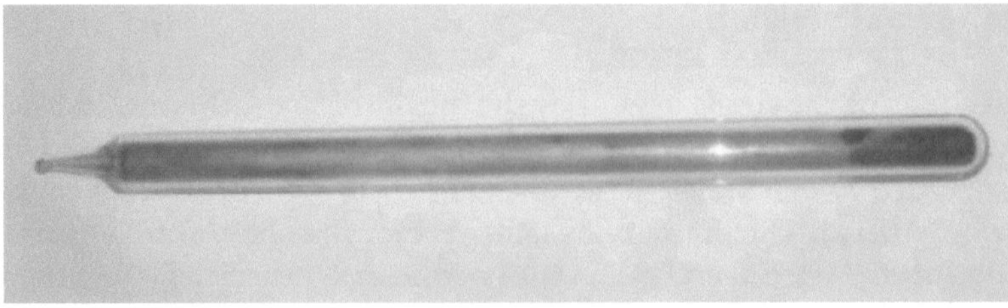

Figure 5.
Sealed glass tube.

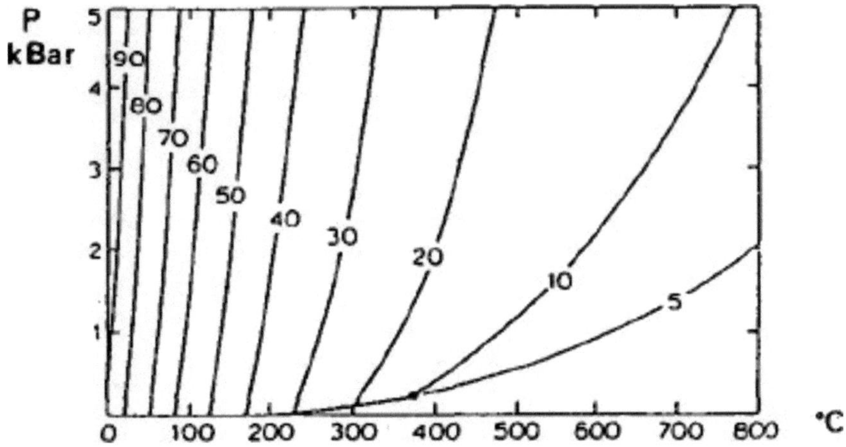

Figure 6.
Variation of the dielectric constant of water as a function of the temperature and the pressure [22].

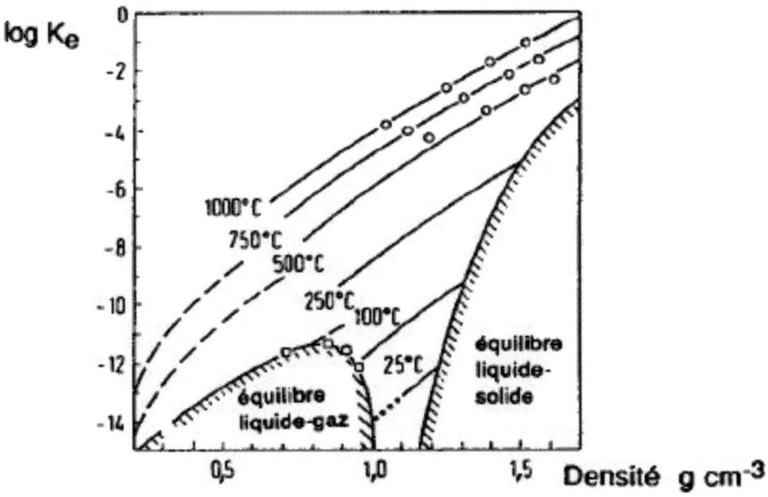

Figure 7.
Variation of the ionic product K_e of water as a function of the temperature and the density of the liquid [24].

The phosphate $AgNi_3PO_4(HPO_4)_2$ [25] has been obtained after 3 weeks of heating at 300°C in sealed glass tube filled with the mixture to about 25% in volume (**Figure 5**). The phase has been prepared from an aqueous solution of $AgNO_3$, $Ni(NO_3)_2.6H_2O$, and H_3PO_4 in the atomic ratio Ag:Ni:P = 2:1:2.

In the other hand, most of the single crystals of the borophosphate family have been obtained by the hydrothermal rout. Kniep et al. [26] have prepared a lot of new borophosphates as a single crystal and they have developed an approach of the

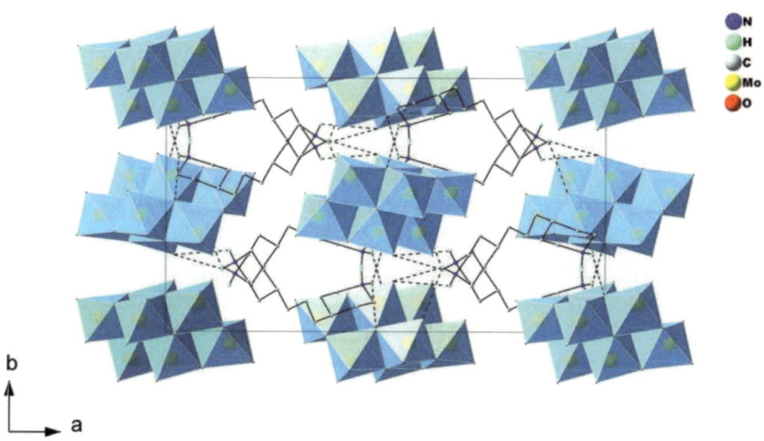

Figure 8.
Projection of the crystal structure of $(C_{13}H_{28}N_2)_2[Mo_8O_{26}]$ *along c-axis [27].*

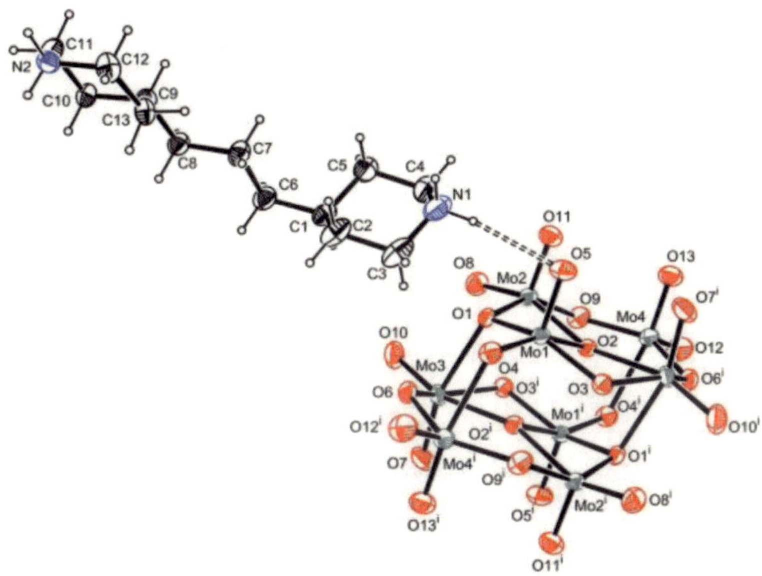

Figure 9.
Representation of the inorganic part $[Mo_8O_{26}]^{4-}$ *and the organic part* $(C_{13}H_{28}N_2)^{2+}$ *of* $(C_{13}H_{28}N_2)_2[Mo_8O_{26}]$ *[27]. The two parts are liked by a hydrogen bond.*

borophosphate crystal chemistry. They have classified the different existing materials in this family as the B/P ratio and as the coordination number of the bore [26].

It is possible to obtained hybrid (organic/inorganic) materials by using this synthesis method. For example, the hybrid material with general formula Bis[4,4′-(propane-1,3-diyl)dipiperidinium]β-octamolybdate (VI) [27] (**Figures 8** and **9**) has been synthesized as single crystals by using the hydrothermal method in an autoclave at 150°C for 2 days.

4. Reaction at high pressure

The effect of pressures on the crystal structure of same materials as the transformation of the ZnO from wurtzite to rock salt from 9 to 13 GPa is well known [28]. Another example is the transformation of the olivine structure at high pressure from the hexagonal close packing into the cubic close packing of the spinel structure [29]. Upon high pressure conditions (6 GPa, 1173 K) olivine-like $LiMAsO_4$ (M = Fe, Co, Ni) transforms to spinel-like compounds where Li^+ and M^{+2} ions randomly occupy 16d octahedral positions and the As^{+5} cations occupy the tetrahedral 8a sites [29] (**Figure 10**).

Since 2006, the prediction of the structure at high pressures became an area of intense activity thanks to the development of the new computer program USPEX [30] by Oganov et al. The code was used with success to predict many new crystal structures, and the results were confirmed by the synthesis of same predicted materials such as Na-Cl system: Na_3Cl, Na_2Cl, Na_3Cl_2, Na_4Cl_3, $NaCl_3$, and $NaCl_7$ [31] and H-Cl system: H_2Cl, H_3Cl, H_5Cl, and H_4Cl_7 [32, 33]. This result allows the discovery of new generation of materials where the core electrons can participate in the formation of chemical bonds. Thus, obviously, we will have very interesting physical and chemical properties.

5. Synthesis by slow evaporation at room temperature

It is the simplest method, but it is suitable only for certain materials. The preparation protocol consists in weighing the reagents in the desired proportion and adding water or an organic base in a ratio that should be determined experimentally. Then the solution must be heated to reach the saturation.

This method takes a few hours to obtain crystals as in the synthesis of $NH_4H_2AsO_4$ or $NH_4H_2PO_4$, and for other materials, it needs a few days. The condensation of ions

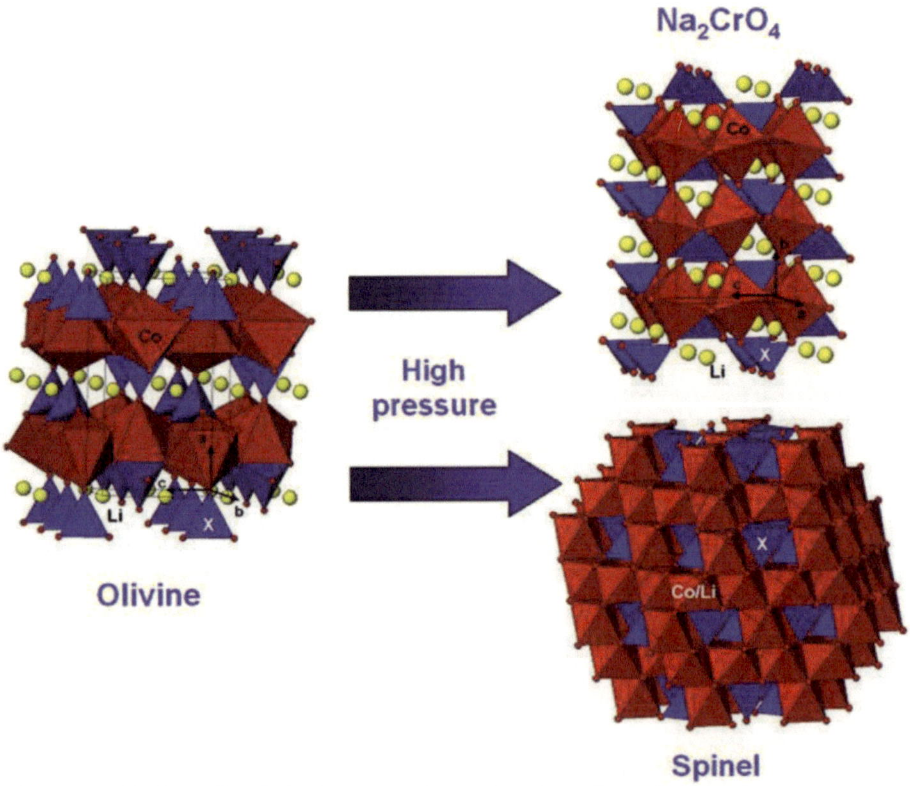

Figure 10.
Structures of olivine-, Na_2CrO_4-, and spinel-like $LiCoXO_4$ (X = P, As) [29].

in aqueous solution forms entities in which cations, identical or different, are linked by different types of oxygenated bridges, HO^- or O^{2-}. Thus, the value of pH is a very important factor in this method. In fact, the choice of pH must be chosen to co-precipitate the various reagents simultaneously. For example, in the ternary system H_2O-NH_3-As_2O_5, the adjusting of the pH at 4.3 allows the formation of $NH_4H_2AsO_4$, and for a basic value of pH, the preparation gives a second phase: $(NH_4)_2HAsO_4$.

The following phosphate $KCdHP_2O_7.2H_2O$ (**Figure 11**) [34] has been obtained by slowly evaporation at room temperature. The method consists in the preparation of a saturated solution containing the reagents in the desired proportion. The reagents are dissolved in water then heated at 100°C for a few minutes to have a saturated solution. The solution obtained is transferred into a petri dish and left in a corner in the laboratory.

The hybrid materials (organic base+ salt) may be synthesized by slow evaporation at room temperature. The best solvent can be used is the water. It is possible to add other solvent with water such as alcohol like in the preparation of $(C_7H_7N_2)_2[CuCl_4].2H_2O$ [35]. In this type of materials, the organic part and the inorganic part are linked via hydrogen bond and π–π bond as shown in **Figure 12**.

Another example may be sited in this section, the synthesis and the crystal structure of $(C_3H_6N_3)_4Bi_2Cl_{10}$ [36] which was grown by a slow evaporation of an aqueous solution of bismuth chloride ($BiCl_3$) and 3-aminopyrazole ($C_3H_5N_3$) in molar amount (1: 2) with a small excess of hydrochloric acid (HCl). The single crystals were obtained after 7 months of slow evaporation (**Figure 13**).

Many organic compounds also have been crystallize by slow evaporation at room temperature such as alkyl-2-(2-imino-4-oxothiazolidin-5-ylidene) acetate [37], 3,4-*cis*-disubstituted pyrrolidin-2-ones [38] ethyl 2-(4-chlorophenyl)-3-cyclopentyl-4-oxo-1-propylimidazolidine-5-carboxylate ($C_{20}H_{27}ClN_2O$) [39]. The obtaining

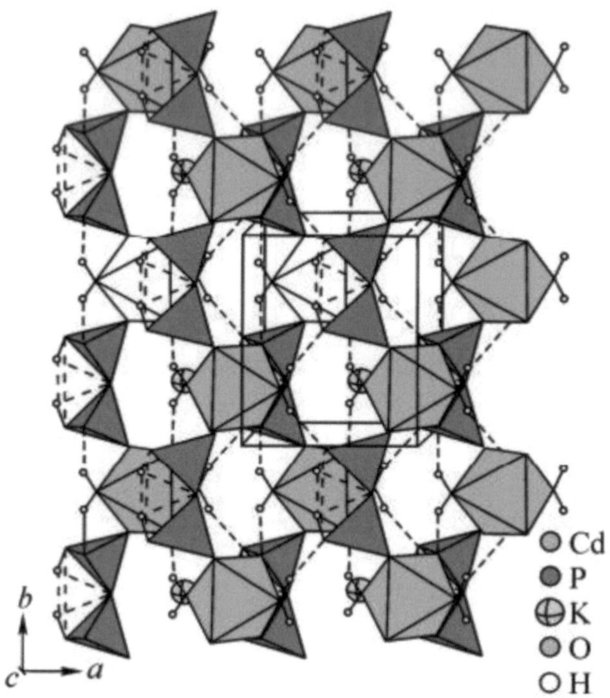

Figure 11.
Perspective view of the structure of $KCdHP_2O_7.2H_2O$ showing the hydrogen bonds (dashed lines) and the location of the alkali metal cations [34].

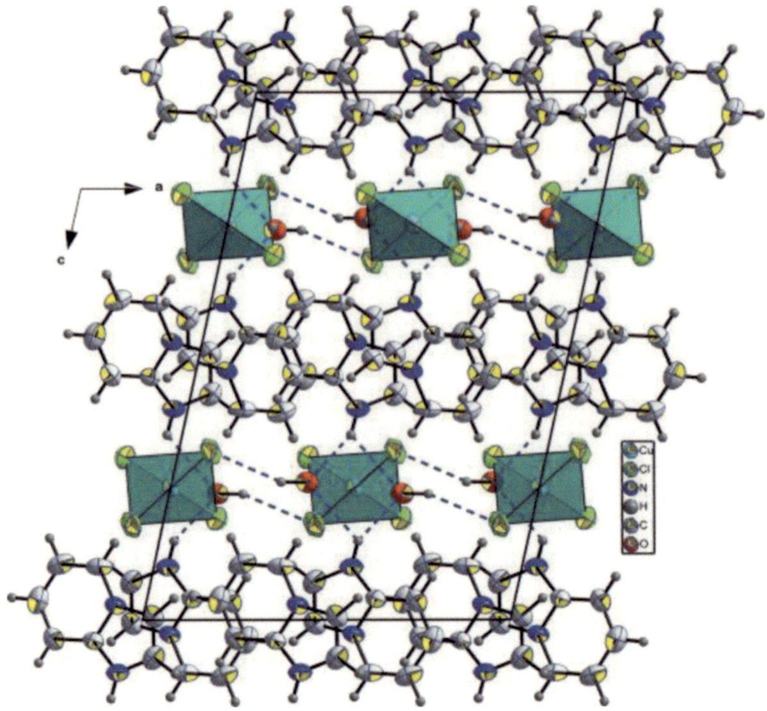

Figure 12.
Projection of the structure of $(C_7H_7N_2)_2[CuCl_4].2H_2O$ showing the alternating stacking of the organic and inorganic layers connected through hydrogen bonds. The face-to-face $\pi-\pi$ stacking between parallel organic molecules is noteworthy with a centroid-centroid distance of 3.968 (3) [35].

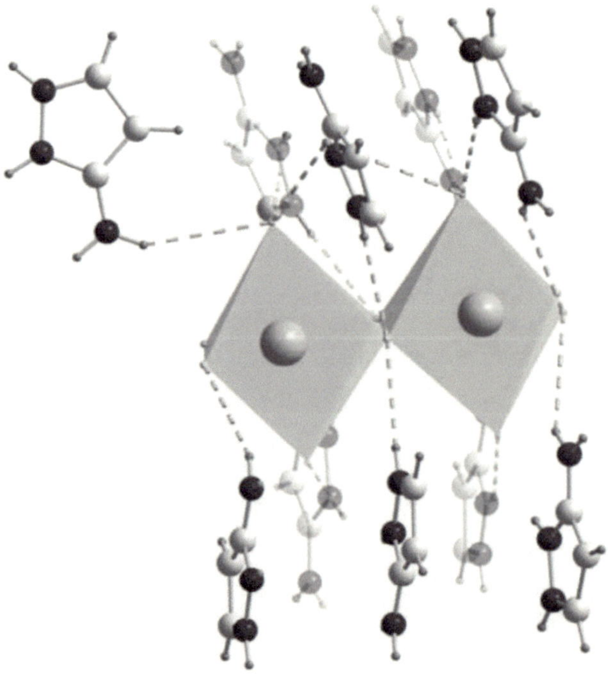

Figure 13.
Bi_2Cl_{10} octahedron surrounded by aminopyrazolium entities, showing the H-Cl contacts [36].

of single crystals depends in the case on the nature of the organic compounds and solvent. The presence of heterocyclic group in the organic molecule promotes the crystallization by the formation of hydrogen and π–π bond.

6. Flux method

Usually, the flux method is used to grow materials as single crystals [40]. The main objective of this method is to decrease the crystallization temperature. This technique has been used to grow high melting phosphate crystals, arsenates, oxides, minerals, and ceramic crystals which cannot be obtained by the conventional solid-state method (Section 1). In this growth technique, the basic materials (solute precursors to crystallize) are reduced to a liquid form in an appropriate flow and the growth process starts when the solution reaches critical supersaturation. The resulting supersaturation and crystal growth are achieved by flow evaporation, solution cooling, or a transport process in which the solute is caused to flow from the hottest region to the coldest region.

The understanding of the phase information about the materials is indispensable to optimize the crystal growth of multi-component system. Practically, binary or other above compositional systems can be easily described using the phase diagram. The phase diagram shows the crystallization or solidification within a material as a function of the material composition (% elements) and material temperature.

The solvent can be a single element, compound, or combination of chemical compounds. The solute is an element or a compound with a melting point generally higher than that of the solvent but in principle it is quite possible to grow crystals from eutectic systems in which the "solvent" has a higher melting point.

For more clarification, taking the example of $LiPr(PO_3)_4$ [40]. The material has been synthesized as a single crystal by the means of flux method. A mixture of Li_2CO_3 and Pr_6O_{11} with stoichiometric ratio was dissolved in an excess of phosphoric acid H_3PO_4 (85%). The mixture was heated to 200°C for 12 h, then to 325°C for 5 days. Finally, the mixture was cooled slowly to room temperature. The single crystal was separated from the excess phosphoric acid by washing in boiling water [40].

7. Crystal growth

When a solid is heated, some physical and chemical changes sintering, melting, and thermal decomposition can be observed. Sintering process results from crystal growth at the contact area between adjacent crystallites. As to conclude, the crystallites connect to each other, and the size increases [41–43].

In high temperatures, as a result of ion movements, melting occurs. The ordered lattice array is replaced by the short-range order of the liquid state. Crystallization may proceed in the light of several different ways:

1. Vapor-solid (condensation)

2. Solution-solid (precipitation)

3. Melt-solid (freezing)

4. Solid A-Solid B (transformation)

It is well-known that there is a remarkable correlation between stability and energy. Stable states have low energy values. For the processes, final energy values of solids must be lower from the free energy of initial state of the systems. It should be noted that crystal formation process forms from two steps. First step is the formation of a new nucleus. Second step is the growth of the nucleus formed to form a particle of appreciable size. Crystals may contain some defects. In terms of the determination of crystal properties, these defects are quite important. In addition to these defects, distortion of lattice also may be possible. This condition is called as dislocation. Many important properties of crystals are due to the regions and numbers of these dislocations. Crystal growths can be via the following types:

- growth from melt
- high-temperature solution growth
- flux growth
- chemical vapor transport
- hydrothermal synthesis
- high-pressure synthesis
- electrolytic reduction of fused salts

In high-temperature solution growth, the constituents of the crystals are dissolved in a suitable solvent and then when the solution becomes over saturated, crystallization occurs. In flux growth, crystals like ceramics and ferrites have been grown by the slow cooling of a solution in a molten flux. Chemical transport reactions are widely considered in the preparation of single crystals like magnetite.

8. Conclusion

In this chapter, the methods of synthesis of crystalline materials and their stages are discussed. These methods are the most adopted and the most common in solid-state chemistry. Each synthesis method brings back to crystalline materials of different sizes returning to the synthesis conditions. The change of one of the parameters such as temperature, pressure, and reactional environment can influence the crystallinity and the size of the sample obtained. For this, the control of these parameters is essential in the synthesis of crystallized materials and in their reproducibility.

Author details

Youssef Ben Smida[1], Riadh Marzouki[2,3,4*], Savaş Kaya[5], Sultan Erkan[6], Mohamed Faouzi Zid[4] and Ahmed Hichem Hamzaoui[1]

1 Laboratory of Valorization of Useful Materials, National Center of Materials Sciences Research, Techno Park Borj Cedria, Carthage University, Soliman, Tunisia

2 Chemistry Department, College of Science, King Khalid University, Abha, Saudi Arabia

3 Chemistry Department, Faculty of Sciences of Sfax, University of Sfax, Tunisia

4 Laboratory of Materials, Crystal Chemistry and Applied Thermodynamics, LR15ES01, Faculty of Sciences of Tunis, University of Tunis El Manar, Tunisia

5 Department of Pharmacy, Health Services Vocational School, Sivas Cumhuriyet University, Sivas, Turkey

6 Yildizeli Vocational School, Chemical and Chemical Company Technology, Cumhuriyet University, Sivas, Turkey

*Address all correspondence to: rmarzouki@kku.edu.sa; riadh.marzouki@hotmail.fr

References

[1] Marzouki R, Sayed AM, Graia M, Zid MF. Cobalt phosphate and application. DOI: 10.5772/intechopen.86215

[2] Ben Smida Y, Guesmi A, Driss A. $LiCo_2As_3O_{10}$: une nouvelle structure à tunnels interconnectés. Acta Crystallographica. 2013;**E69**:i39

[3] Ben Smida Y, Marzouki R, Guesmi A, Georges S, Zid MF. Synthesis, structural and electrical properties of a new cobalt arsenate $NaCo_2As_3O_{10}$. Journal of Solid State Chemistry. 2015;**221**:132-139

[4] Ben Smida Y, Marzouki R, Georges S, Kutteh R, Avdeev M, Guesmi A, et al. Synthesis, crystal structure, electrical properties, and sodium transport pathways of the new arsenate $Na_4Co_7(AsO_4)_6$. Journal of Solid State Chemistry. 2016;**239**:8-16

[5] ALQarni OSA, Marzouki R, Ben Smida Y, Alghamdi MM, Avdeev M, Belhadj Tahar R, et al. Synthesis, electrical properties and Na^+ migration pathways of $Na_2CuP_{1.5}As_{0.5}O_7$. Processes; **8**(3):305

[6] Ouerfelli N, Ben Smida Y, Zid MF. Ac conductivity, electric modulus analysis, dielectric behavior and bond valence sum analysis of $Na_3Nb_4As_3O_{19}$ compound. Journal of Alloys and Compounds. 2015;**651**:616-622

[7] Marzouki R, Ben Smida Y, Guesmi A, Georges S, Ali IH, Adams S, et al. Structural and electrical investigation of new melilite compound $K_{0.86}Na_{1.14}CoP_2O_7$. International Journal of Electrochemical Science. 2018;**13**:11648-11662

[8] Marzouki R, Ben Smida Y, Sonni M, Avdeev M, Zid MF. Synthesis, structure, electrical properties and Na^+ migration pathways of $Na_2CoP_{1.5}As_{0.5}O_7$. Journal of Solid State Chemistry. 2020;**285**:121058

[9] Marzouki R, Ben Smida Y, Avdeev M, Alghamdi MM, Zid MF. Synthesis, structure and Na^+ migration pathways of new Wylleite-type $Na_{1.25}Co_{2.187}Al_{1.125}(AsO_4)_3$. Materials Research Express. 2019;**6**:126313

[10] Ben Smida Y, Guesmi A, Zid MF, Driss A. $Na_3Co_2(As_{0.52}P_{0.48})O_4(As_{0.95}P_{0.05})_2O_7$. Acta Crystallographica. 2013;**E69**:I85-I86

[11] Falah C, Ben Smida Y, Ledoux-Rak I, Boughzala H. Synthesis, crystal structure and ionic conductivity of a new open-framework arsenate $K_{0.40}5Bi_{0.865}AsO_4$. Journal of Alloys and Compounds; **653**:321-326

[12] Marzouki R, Guesmi A, Zid MF, Driss A. Synthesis, crystal structure and electrical properties of a new mixed compound $(Na_{0.71}Ag_{0.29})_2CoP_2O_7$. Crystal Structure Theory and Applications. 2012;**1**:68-73

[13] Marzouki R, Guesmi A, Georges S, Zid MF, Driss A. Structure, sintering and electrical properties of new ionic conductor $Ag_4Co_7(AsO_4)_6$. Journal of Alloys and Compounds. 2014;**586**:74-79

[14] Marzouki R. Electrical properties and alkali-pathways simulation of new mixed conductor $Na_4Li_{0.62}Co_{5.67}Al_{0.71}(AsO_4)_6$. Materials Research Express; **7**:16313

[15] Ben Moussa MA, Marzouki R, Brahmia A, Georges S, Obbade S, Zid MF. Synthesis and structure of new mixed silver cobalt(II)/(III) diphosphate - $Ag_{3.68}Co_2(P_2O_7)_2$. Silver(I) transport in the crystal. International Journal of Electrochemical Science. 2019;**14**:1500-1515

[16] Nasri R, Marzouki R, Georges S, Obbade S, Zid MF. Synthesis, sintering, electrical properties and sodium migration pathways of new lyonsite

$Na_2Co_2(MoO_4)_3$. Turkish Journal of Chemistry. 2018;**42**:1251-1264

[17] Marzouki R, Zid MF. Non-centrosymmetric $Na_7Li_{0.8}K_{0.2}Co_5(As_3O_{10})_2(As_2O_7)_2$: Synthesis, structure and alkali ion-conduction pathways simulation. International Journal of Electrochemical Science. 2020;**15**:3776-3792

[18] Rabenau A. The role of hydrothermal synthesis in preparative chemistry. Angewandte Chemie (International Edition in English). 1985;**24**:1026

[19] Stein A, Keller SW, Mallouk TE. Turning down the heat: Design and mechanism in solid-state synthesis. Science. 1993;**259**:1558

[20] Barrer RM. Hydrothermal Chemistry of Zeolites. London: Academic Press; 1982

[21] Haar L, Gallagher JS. NBSBVRC Steam Tables. Hemisphere Publ.; 1984

[22] Rickard D, Wickman E, editors. Physics and Chemistry of the Earth. New York: Pergamon Press; 1981. pp. 113-132

[23] Todheide K. In: Francks E, editor. Water, a Comprehensive Treatise. Vol. 1. New York: Plenum; 1972. pp. 463-514

[24] Holzapfel W, Franck EU. Leitfiihigkeit und lonendissoziation des Wassers bei 1000°C und 100 kbar. Berichte der Bunsengesellschaft für Physikalische Chemie. 1966;**70**:1105

[25] Ben Smail R, Juini T. $AgNi_3(PO_4)(HPO_4)_2$: An alluaudite-like structure. Acta Crystallographica. 2002;**C58**:i61-i62

[26] Kniep R, Engelhardt H, Hauf C. A first approach to borophosphate structural chemistry. Chemistry of Materials. 1998;**10**:2930-2934

[27] Driss M, Ksiksi R, Ben Amor F, Zid MF. Bis[4,40-(propane-1,3-diyl) dipiperidinium] b-octamolybdate(VI). Acta Crystallographica. 2010;**E66**:m533-m534

[28] Thiyagarajan R, Yan X, Pazhanivelu V, Selvadurai APB, Murugaraj R, Yang W. Doping effect of alkali metal elements on the structural stability and transport properties of ZnO at high pressures. Journal of Alloys and Compounds. 2018;**751**:266-274

[29] Amador U, Gallardo-Amores JM, Heymann G, Huppertz H, Moran E, Arroyo y de Dompablo ME. High pressure polymorphs of $LiCoPO_4$ and $LiCoAsO_4$. Solid State Sciences. 2009;**11**:343-348

[30] Oganov AR, Glass CW. Crystal structure prediction using ab initio evolutionary techniques: Principles and applications. The Journal of Chemical Physics. 2006;**124**:244704

[31] Zhang W, Oganov AR, Goncharov AF, Zhu Q, Boulfelfel SE, Lyakhov AO, et al. Unexpected stable stoichiometries of sodium chlorides. Science. 2013;**342**:1502

[32] Zeng Q, Yu S, Li D, Oganov AR, Frapper G. Emergence of novel hydrogen chlorides under high pressure. Physical Chemistry Chemical Physics. 2017;**19**:8236

[33] Duan D, Huang X, Tian F, Liu Y, Li D, Yu H, et al. Predicted formation of H^{3+} in solid halogen polyhydrides at high pressures. The Journal of Physical Chemistry A. 2015;**119**:11059-11065

[34] Kouass S, Selmi A, Chebbi H, Guesmi A. Characterization and crystal structure of a new layered cadmium diphosphate: $KCdHP_2O_7{\cdot}2H_2O$. Journal of Structural Chemistry. 2016;**57**:679-683

[35] Mokaddem S, Boughzala H. Bis(imidazo[1,2-a]pyridin-1-ium)

tetrachloridocuprate(II) dihydrate. Acta Crystallographica. 2017;**E73**:196-199

[36] Ferjani H, Boughzala H. New hybrid material: $(C_3H_6N_3)_4Bi_2Cl_{10}$. Synthesis, structural study and spectroscopic behavior. Russian Journal of Inorganic Chemistry. 2018;**63**:349-356

[37] Sbei N, Haouas B, Chebbi M, Ben Smida Y, Arfaoui Y, Boujlel K, et al. A convenient synthesis of alkyl-2-(2-imino-4-oxothiazolidin-5-ylidene) acetate derivatives involving an electrogenerated base of acetonitrile. Journal of Sulfur Chemistry. 2017;**38**:152-162

[38] Arfaoui A, Saadi F, Ben Smida Y, Arfaoui Y, Nefzi A, Amri H. A convenient synthesis of 3, 4-cis-disubstituted pyrrolidin-2-ones. Tetrahedron Letters. 2015;**56**:98-100

[39] Tabarki MA, Ben Smida Y, Guesmi A, Besbes R. Crystal structure of ethyl 2-(4-chlorophenyl)-3-cyclopentyl-4-oxo-1-propylimidazolidine-5-carboxylate. Acta Crystallographica. 2015;**E71**:0682-0683

[40] Gharouel S, Smida YB, Ferhi M, Marin JI, Romo LA, Pablo KH-N, et al. Synthesis, structural, electrical and optical properties of $LiPr(PO_3)_4$. Journal of Solid State Chemistry. 2020;**290**:121459

[41] Verneuil MA, Acad CR. Note sur Ie parallésme des roches des dépôts paléozoïques de l'Amérique Septentrionale avec ceux de l 'Europe. Comptes rendus de l'Académie Sciences Paris. 1902;**135**:791

[42] Remeika PJ, Jackson WM. A method for growing barium titanate single crystals. Journal of the American Chemical Society. 1954;**76**:940

[43] Wold VA, Dwight K. Solid State Chemistry. Synthesis, Structure, and Properties of Selected Oxides and Sulfides. New York: Chapman & Hall; 1993. p. 245

Chapter 3

One-Pot Synthesis of Chiral Organometallic Complexes

Mei Luo

Abstract

Currently, organometallic complexes involving ligand oxazolines are typically obtained in two sequential steps, where the free ligand is given firstly from a functionalized nitrile by condensation reaction with an amino alcohol in the presence of a Lewis or Bronsted acid catalyst, followed by a further coordination with metal salts to obtain the corresponding oxazolinyl metal complexes. Usually, the yield of the two-step procedure is relatively low; considering that metal oxazoline complexes often contain Lewis acidic metals, it is possible that the two steps may be telescoped. A series of novel chiral organometallic complexes (**1–23**) were assembled in a single step from nitriles, chiral D/L amino alcohols, and a stoichiometric amount of metal salts ($MCl_2{\cdot}nH_2O/M(OAc)_2{\cdot}nH_2O$), with moderate to high yields (20–95%). All the crystalline compounds were fully characterized by NMR, IR, MS, and X-ray analyses.

Keywords: chiral organometallic complexes, nitriles, amino alcohols, metal salts, crystalline compounds

1. Introduction

Chiral oxazolines constitute an important class of "privileged" ligands in asymmetric catalysis [1–8]. Organometallic complexes involving oxazoline ligands are typically obtained in two steps, where the free ligand is given firstly from a functionalized nitrile through condensation reaction with an amino alcohol in the presence of a Lewis or Bronsted acid catalyst, followed by further complexed coordination with metal salts to obtain the corresponding oxazolinyl metal complexes (**Figure 1**) [9, 10]. Usually, the yield of the two-step procedure is relatively low, and certain oxazolinyl organometallic complexes are difficult to obtain due to the poor coordination ability of the imine group from the oxazoline. It is conceivable that the two steps may be telescoped by using the requisite Lewis acid precursor. Herein, through the assembly of three reaction components (a nitrile, an amino alcohol, and metal salts), we first report a simple, one-step procedure for the preparation of N-containing heterocyclic zinc complexes (**1–15**), with the yield of certain products reaching 90% in the presence of a large amount of $ZnCl_2$ (0.4–2.6 eq.) and certain chiral salicyloxazoline metal complexes (**16–23**) with yields ranging from 65 to 95% using 1.0 eq. of copper, cobalt, nickel, manganese, palladium, and platinum salts as the third component. In all the cases, the complexes were isolated, purified, and characterized. All the structures reported in this paper were confirmed by X-ray crystallography.

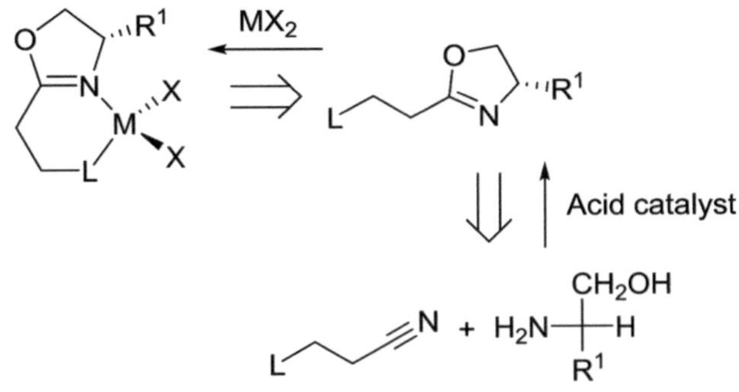

Figure 1.
Common method for the preparation of oxazolinyl metal complexes.

2. One-step multicomponent synthesis of chiral oxazolinyl-zinc complexes

The one-pot procedure was initially tested from the reaction of different 1-piperidine propionitrile derivatives with 2–3 eq. of amino alcohol refluxed in chlorobenzene for 72 h in the presence of 1–2.6 eq. of $ZnCl_2$. After cooling to room temperature, the solvent was removed under reduced pressure, and the residue was dissolved in H2O and extracted with CH_2Cl_2. The combined organic extracts were evaporated to give a crude red oil, which was purified by column chromatography (petroleum ether/CH_2Cl_2, 4/1) to afford the title compound as crystals. During the preliminary work, it quickly became apparent that the reaction results are controlled by the amount of $ZnCl_2$ used (**Figure 2**); for example, employing 1.1 eq. or 2.6 eq. of $ZnCl_2$, the desired crystal structures of the amino-oxazolidinyl zinc complex **1** and bis-oxazolidinyl zinc complex **2** containing two monodentate ligands can be obtained from the reaction of L-leucinol or L-valinol with 3-piperidin-1-yl-propionitrile, respectively, followed by evaporation of different ratios of petroleum and dichloromethane from the mixture after column separation but with only a low yield (25%) for complex **1** and moderate yield (65%) for complex **2**.

The nature of the side chain (R^1) influenced the reaction outcome. Using L-phenylalaninol with 1.6 eq. of $ZnCl_2$ or 1.5 eq. of $ZnCl_2$ with 1-morpholinepropionitrile (X = O) and D-phenylglycinol, both led to the cleavage of the propionitrile, providing asymmetric diamine complexes **3** and **4** at very good yields of 86% and 90%, respectively. Interestingly, using 1-(2-cyanoethyl)-4-methylpiperazine (Z = NMe) as a precursor with 2.5 eq. of $ZnCl_2$ led only to the formation of the zwitterionic piperazine complex **5**, irrespective of the amino alcohol used. The results again prove the effects of different amounts of metal salts on the reaction.

From the crystal structures of complexes **2–5**, we conclude that the propionitrile precursors are unstable and decompose into acetonitrile or the parent cyclic amines to afford complexes **2** and **3–5**, respectively, in the presence of a large amount of zinc chloride. For this reason, a number of nitrile precursors with additional N-donor were selected to be more robust against degradation under the reaction conditions. Consequently, a number of aromatic nitrile precursors containing additional N-donors were applied widely in these three-component reactions. In the process of selecting these reactions, the appropriate amount of $ZnCl_2$ was carefully optimized to ensure specific results. Complex **6**, listed in **Figure 3**, contains two monodentate ligands coordinated via the oxazoline nitrogen and was afforded from the use of 3-aminobenzonitrile and D-leucinol in the presence of 0.44 eq.

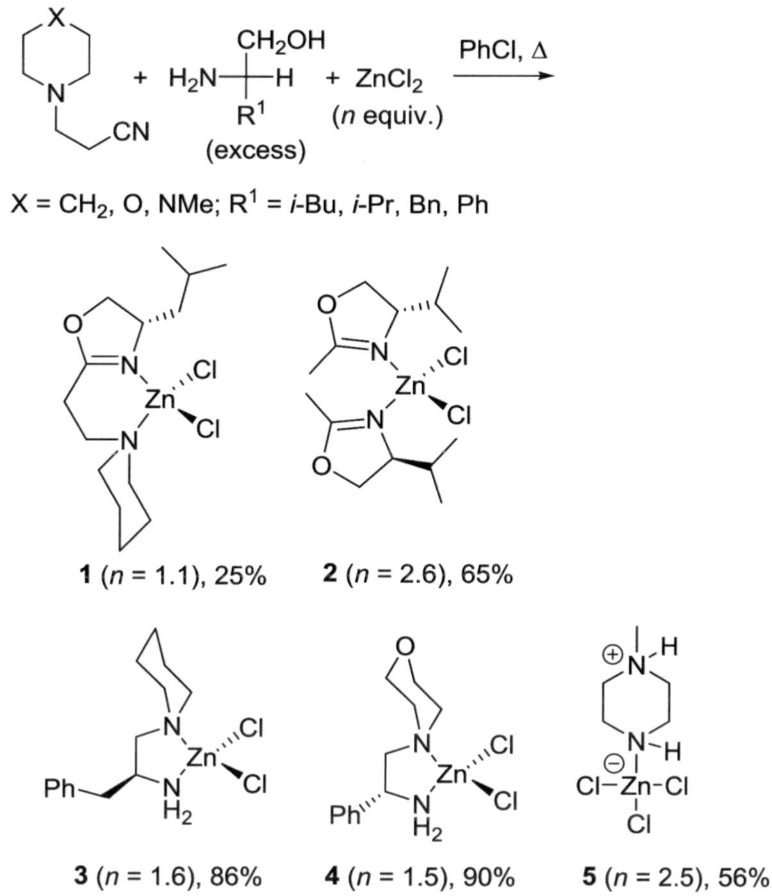

Figure 2.
Effect of the reaction stoichiometry of (metal precursor) $ZnCl_2$.

of $ZnCl_2$. Similarly, the bis-chelated complex 7 and the mono-chelated complex **8** were afforded to the corresponding yields of 80 and 78% from the use of 2-cyano-pyridine with L-phenylalaninol and D-valinol, respectively, in the presence of 1.2 eq. of $ZnCl_2$.

The formation of complexes **9** and **10** were studied using different amino alcohols, as seen in **Figure 4**. C2-symmetrical bis-oxazolines formed seven-membered chelate rings which derived from 1,2-dicyanobenzene, affording a 1:1 adduct with zinc dichloride. Indeed, the addition of isophthalonitrile with D-phenylglycinol (0.56 eq.) provided the predicted mono-chelated complex **9** [11] at a good yield (68%). However, a combination of a slight excess of L-valinol (0.72 eq.) caused the addition of three amino alcohols to give complex **10** with a yield of 66%.

Surprisingly, the combination of L-leucinol and L-phenylglycinol with tetra-cyanoethylene in the presence of 0.42 eq. of $ZnCl_2$, respectively, provided neutral bis[bis(oxazoline)]zinc (II) complexes **11** and **12** with corresponding yields of 88 and 86% (**Figure 5**). The crystal structures of these methylene-bis(oxazoline) indicate that the tricyanomethane was formed as an intermediate from a disproportionation-rearrangement of the tetracyanoethylene precursor, although the precise mechanism of this pathway is unclear. In 2016, Kögel et al. reported the synthesis of complex **12** by a different route [12]. Interestingly, complex **12** was reported to exhibit an intense cotton effect as a result of exciton coupling. Indeed, the X-ray

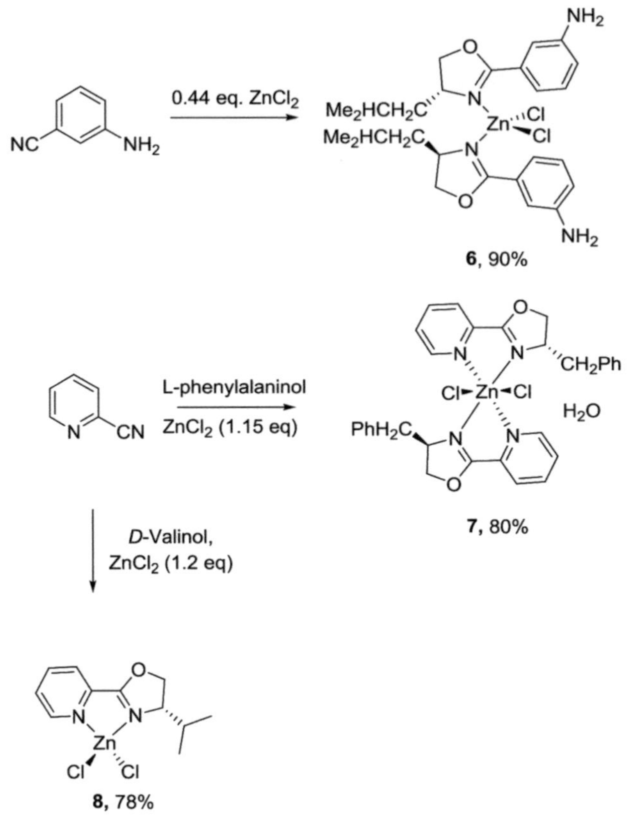

Figure 3.
*Zinc complexes **6–8** derived from 3-aminobenzonitrile and 2-cyanopyridine.*

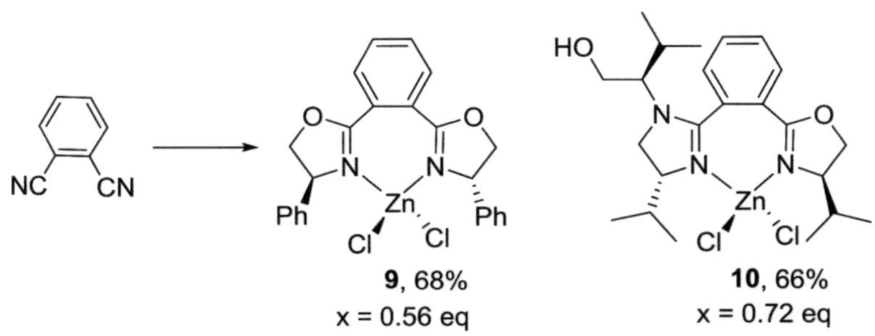

Figure 4.
Complexes 9–10 derived from isophthalonitrile.

crystal structures of complexes **11** and **12** have been proven that due to the their coordination environments, isobutyl-substituted complex **11** has shown a fairly symmetrical tetrahedral comformation, while complex **12** is in highly distorted. This maybe the result of the favorable intramolecular π-interaction between one of the phenyl groups with the semicorrin structure of the adjacent ligand within 3.5 Å, effectively leading to the two chiral chromophores' close proximity to convenience exciton coupling [13].

In further study, 2-hydroxy-6-methylnicotinonitrile was employed as a precursor to test the applicability of the one-pot methodology in assembling complex multinuclear structures. In the presence of different amounts of $ZnCl_2$ (1.72, 1.31,

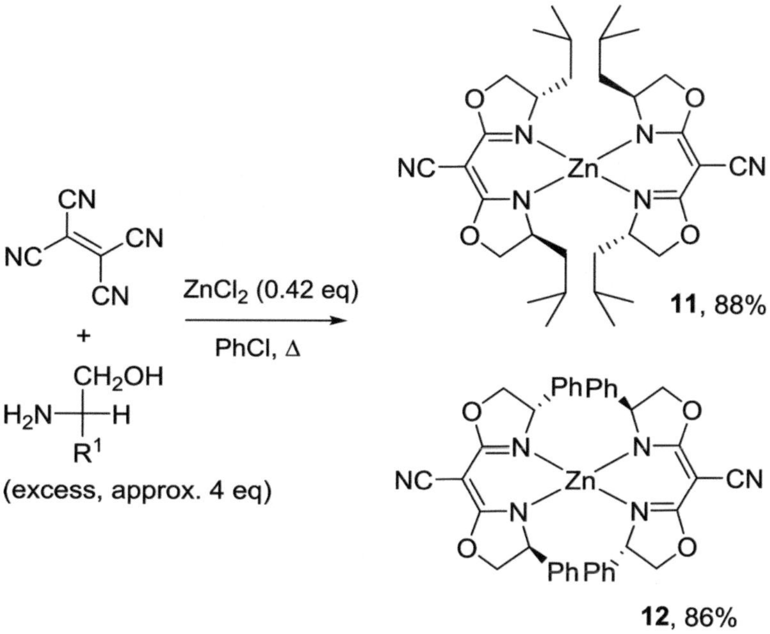

Figure 5.
Neutral zinc complexes derived from tetracyanoethylene.

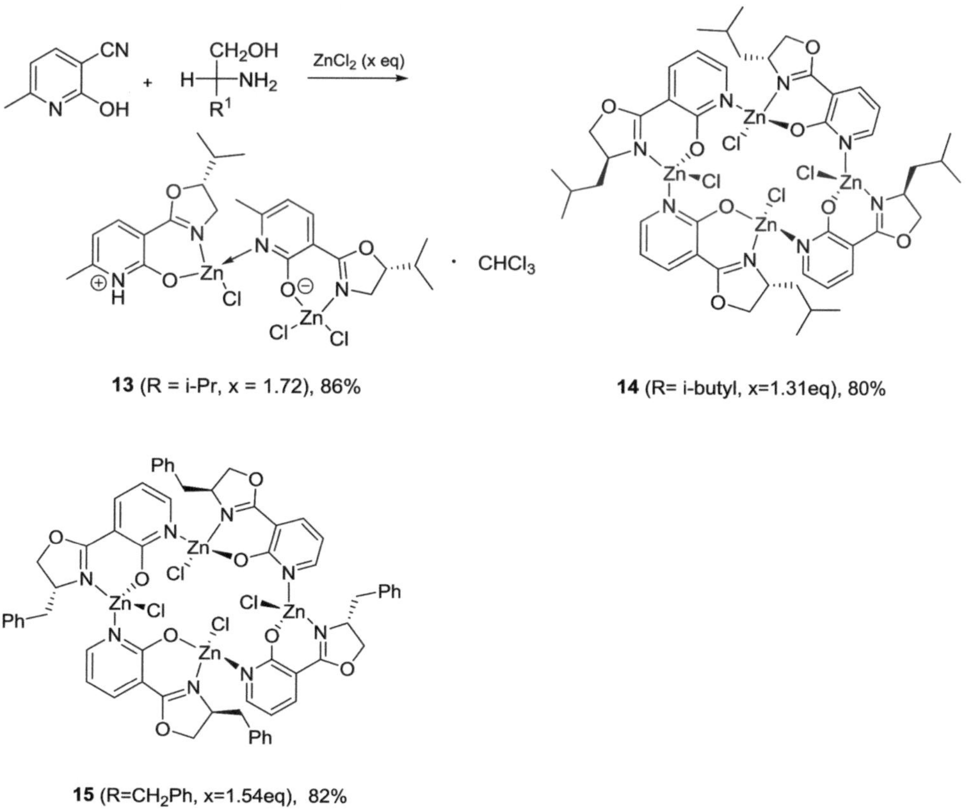

Figure 6.
*Multinuclear zinc complexes **13–15**.*

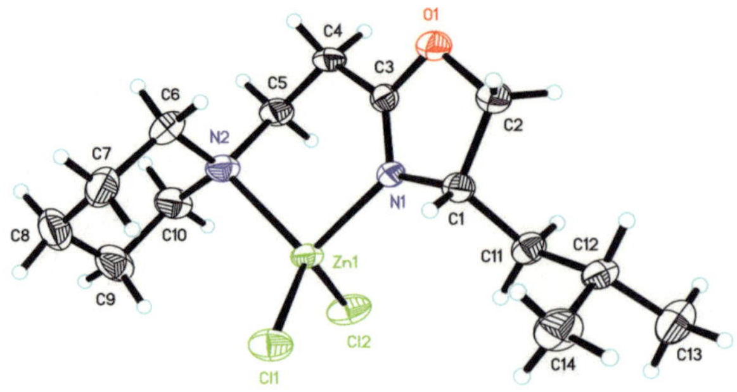

Figure 7.
The crystal structure of complex 1.

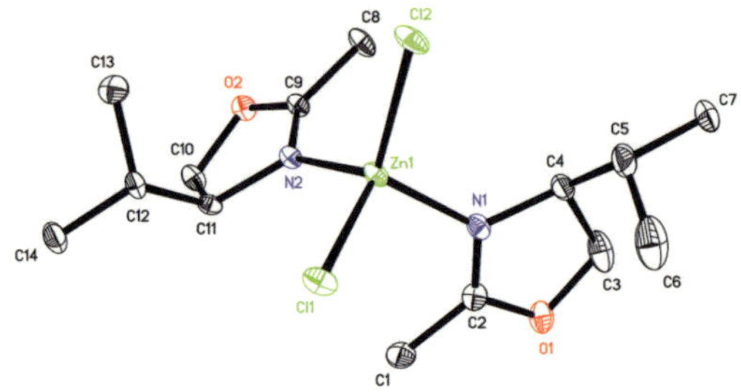

Figure 8.
The crystal structure of complex 2.

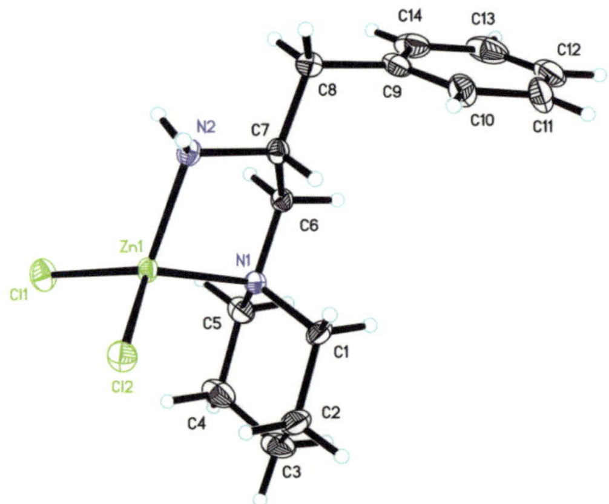

Figure 9.
The crystal structure of complex 3.

and 1.54 eq.), the corresponding condensation products with valinol, leucinol, or phenylalaninol furnished the binuclear zwitterionic complex **13** and highly symmetrical tetramers **14** and **15** at yields of 86, 80, and 82%, accordingly (**Figure 6**). Presumably, the formation of higher aggregates is prevented by the sterically demanding isopropyl substituents. A six-membered N,O-chelate ligand is complexed at each zinc metal center and connected to another metal center with a bridging donor ligand from the pendant pyridine. With each zinc atom located at a corner of a square grid, the planar N,O,N-ligands are oriented perpendicularly to one another with diagonal Zn···Zn distances of ca. 6 Å.

The crystal structures of all the complexes (**Figures 7–21**) are determined and reported by X-ray diffraction, elemental analysis, and IR. In all the cases, a distorted tetrahedral geometry is found at zinc(II), and the C=N double-bond character of the oxazolinyl ligand is largely retained in the metal complexes (**Table 1**).

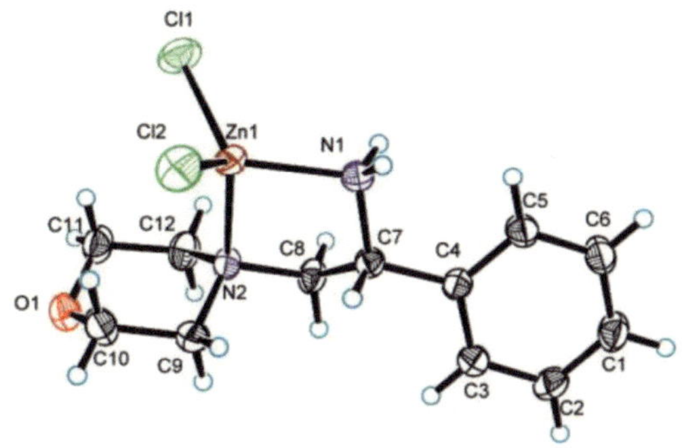

Figure 10.
The crystal structure of complex 4.

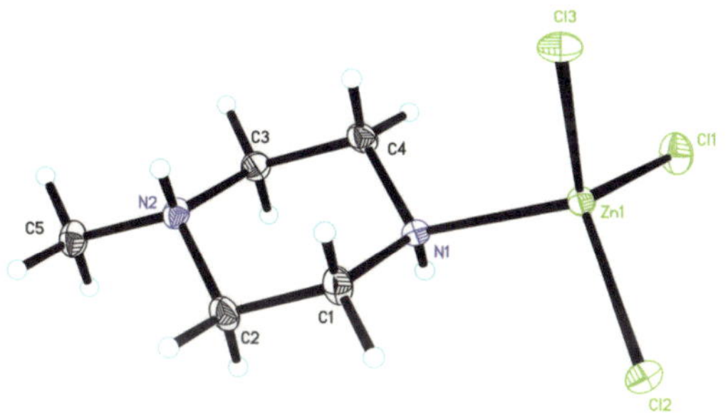

Figure 11.
The crystal structure of complex 5.

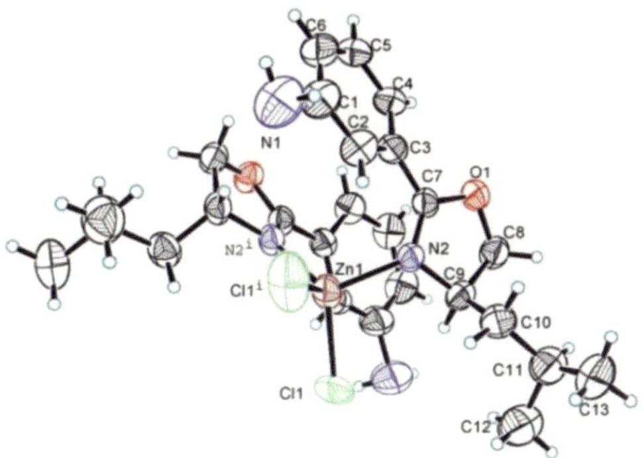

Figure 12.
The crystal structure of complex 6.

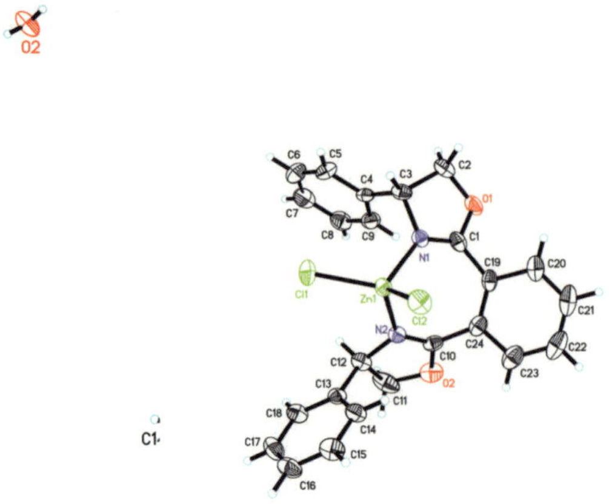

Figure 13.
The crystal structure of complex 7.

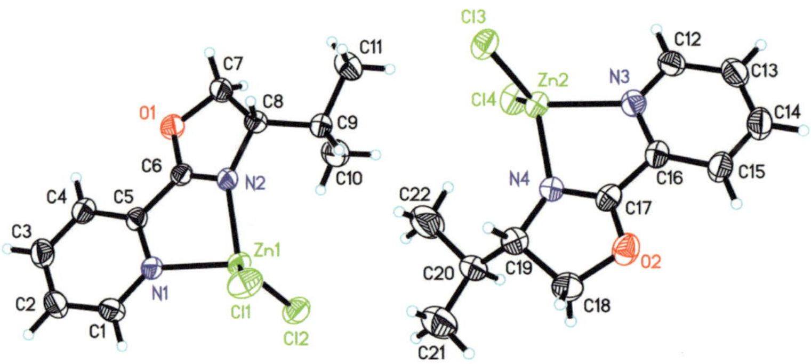

Figure 14.
The crystal structure of complex 8.

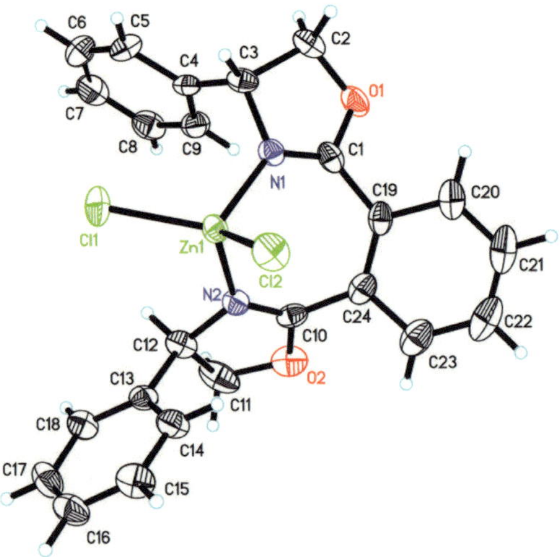

Figure 15.
The crystal structure of complex 9.

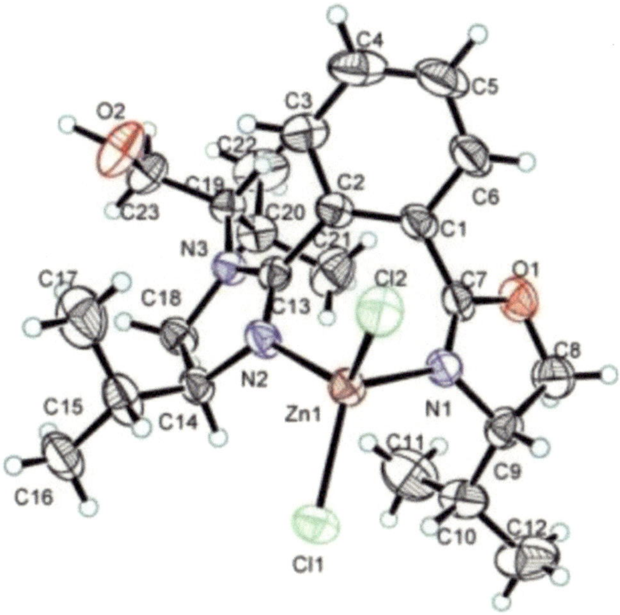

Figure 16.
The crystal structure of complex 10.

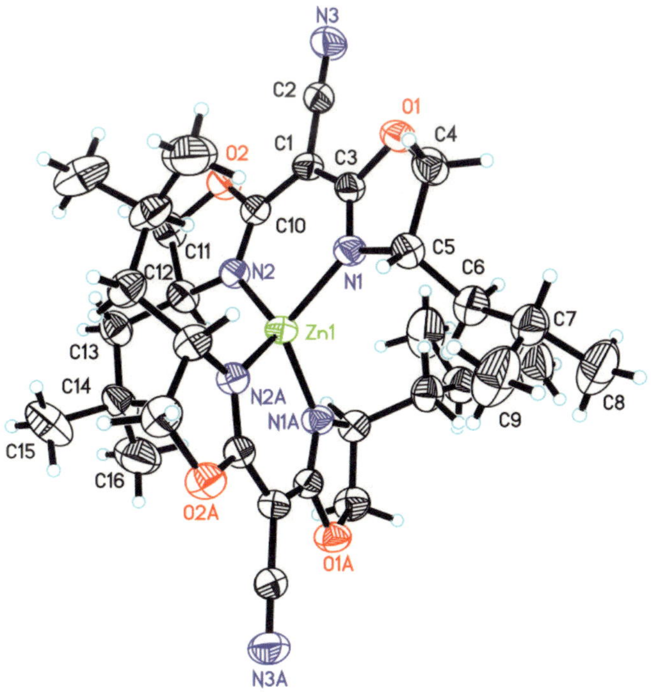

Figure 17.
The crystal structure of complex 11.

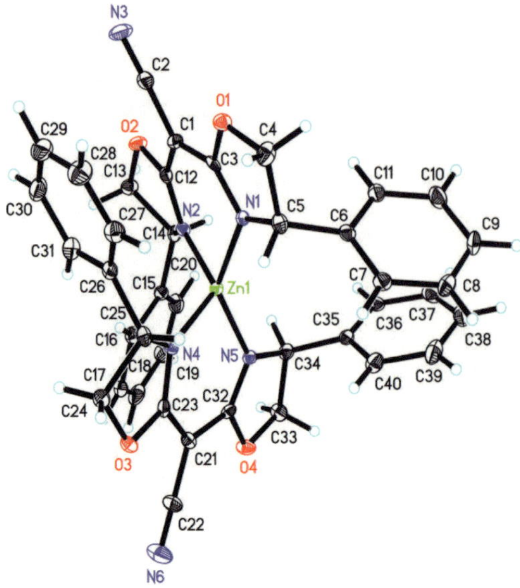

Figure 18.
The crystal structure of complex 12.

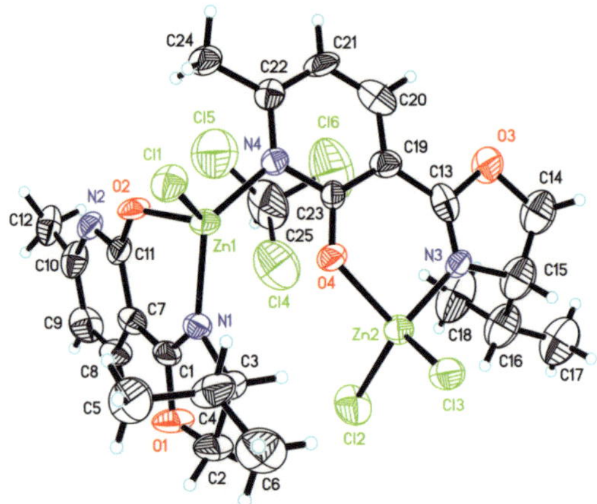

Figure 19.
The crystal structure of complex 13.

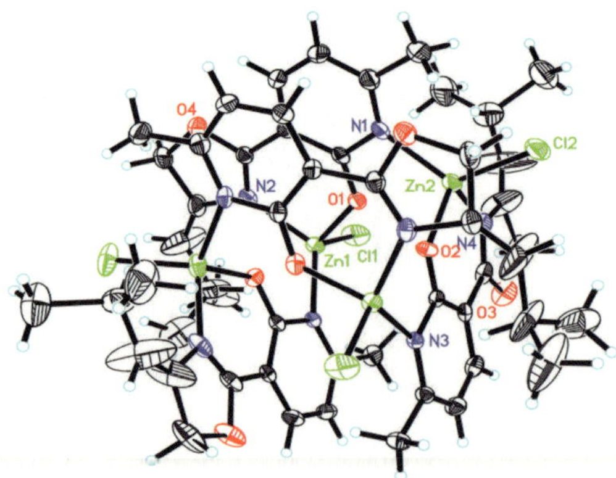

Figure 20.
The crystal structure of complex 14.

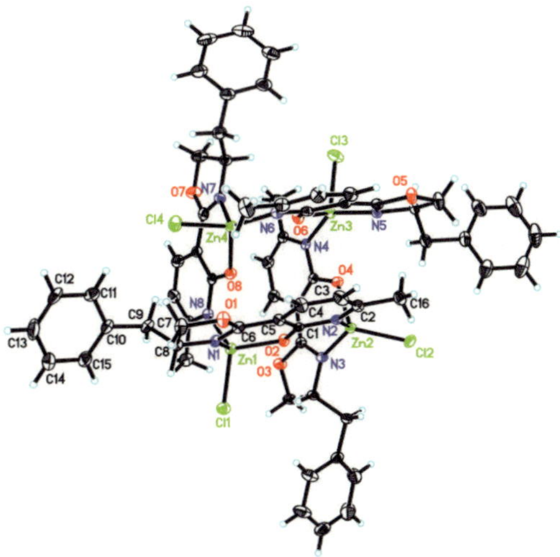

Figure 21.
The crystal structure of complex 15.

$ZnCl_2$ (%)	Products	Yield (%)[a]
114.2	1	25
259.8	2	65
152.8	3	86
144.6	4	90
245.8–256.0	5	56
44.1	6	90
121.8	7	80
122.9	8	85
56.1	9	86
72.6	10	90
42.2	11	88
42.2	12	86
172.1	13	86
130.7	14	80
153.6	15	82`

[a]*Isolated yield from silica gel.*

Table 1.
One-pot synthesis of zinc complexes **(1–15)**.

3. One-step templated synthesis of chiral organometallic salicyloxazoline complexes

Chiral oxazolinyl organometallic complexes are very important catalysts in organic chemistry [14–22]. Several organometallic complexes containing 2-(2′-hydroxyphenyl) oxazolines are reported in the literature [23–42]. The general approach to the synthesis of metal complexes begins with ligand synthesis, followed by ligand reaction with metal salts to afford organometallic complexes [43].

In 2017, our research group first reported a one-pot multicomponent synthesis of chiral oxazolinyl-zinc complexes [44] in the presence of a large amount of $ZnCl_2$ (0.4–2.6 eq.); the yields of certain products reached 90%. Herein, the chiral salicyloxazoline metal complexes **16–23** can be obtained by using 1.0 eq. of metal salts such as copper, cobalt, nickel, manganese, palladium, and platinum salts as the third component. The structures of these complexes were characterized by X-ray crystallography. The results prove that organometallic complexes can be assembled with two reactants and different amounts of metal salts.

Chiral bis(oxazoline) copper complex **16**, nickel complex **17**, cobalt complex **18**, and palladium complex **19** were generated as crystals with the chemical formula ML_2 (L = 2-(4-R_1–4,5-dihydrooxazol-2-yl)phenol, R_1: D-Ph, M: Cu, Ni, Co; R_1: L-CH_2Ph; M: Pd). The syntheses of these complexes can be summarized as follows: A mixture of 2-hydroxybenzonitrile and D-phenylglycinol or L-phenylalaninol in 50 mL of chlorobenzene was refluxed for 72 h with 1.0 eq. of each of the above appropriate metal salts. After removal of the chlorobenzene, single crystals of chiral bis(oxazolinyl) metal complexes **16–19** were present after natural evaporation of the recrystallization or chromatographic solvent with petroleum and dichloromethane (**Figure 22**).

In **Figures 23** and **24**, refluxing a mixture of 2-cyanophenol and d-phenylglycinol in chlorobenzene for 72 h with 1.0 eq. of cobalt chloride hexahydrate or 1.0 eq. of cobalt acetate tetrahydrate, respectively, afforded complexes **20** and **21**. Further, through slow evaporation from a 1:1 mixture of ethanol and chloroform, crystals of complex **20** were obtained. However, the crystals of complex **21** were present after column separation with a 4:1 solution of petroleum ether and dichloromethane, followed by evaporation of the volatile components.

Notably, the product complexes **18** and **20** were obtained using $CoCl_2{\cdot}6H_2O$ as a reagent with different solvents in the workup procedure. As seen in **Figure 23**, when a nonpolar solvent, such as petroleum ether or n-hexane, was used in the recrystallization medium, crystals of complex **18** were obtained. However, if the recrystallization was carried out with a mixture of two polar solvents, such as ethanol and chloroform, crystals of complex **20** were obtained.

Similarly, in the synthesis of chiral oxazoline manganese complex **22** by the title method, 2-hydroxybenzonitrile and D-phenylglycinol were refluxed with 1.0 eq. of manganese acetate tetrahydrate in chlorobenzene for 60 h (**Figure 24**). After removal of the chlorobenzene and slow evaporation with a mixture of absolute ethanol and chloroform, crystals of complex **22** were obtained.

Interestingly, in **Figure 25**, when employed by 1.0 eq. of PtCl2 in the reaction of 2-hydroxybenzonitrile with D-phenylglycinol in chlorobenzene, the crystal structure of the resulting Pt complex was different from the aforementioned complexes **16–22** (**Figures 26–32**); a complex containing

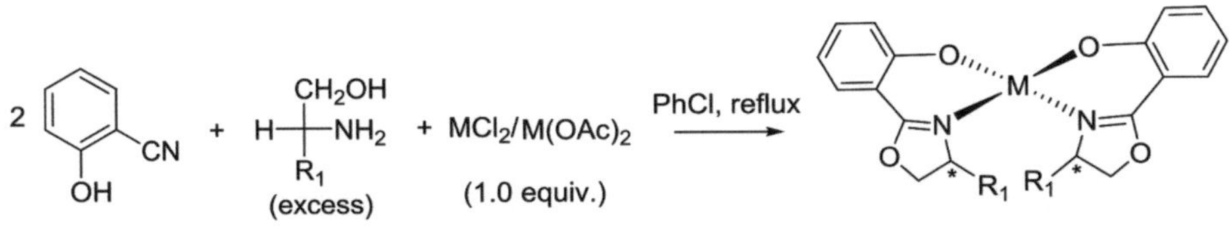

16,17,18 R_1:D-Ph
16: $Cu(OAc)_2$, 65% or $CuCl_2$, 85%
17: $Ni(OAc)_2$, 92% or $NiCl_2$, 95%
18: $CoCl_2$, 72%
19: R_1:L-CH_2Ph; $PdCl_2$, 86%

Figure 22.
Templated synthesis of complexes ***16–19****.*

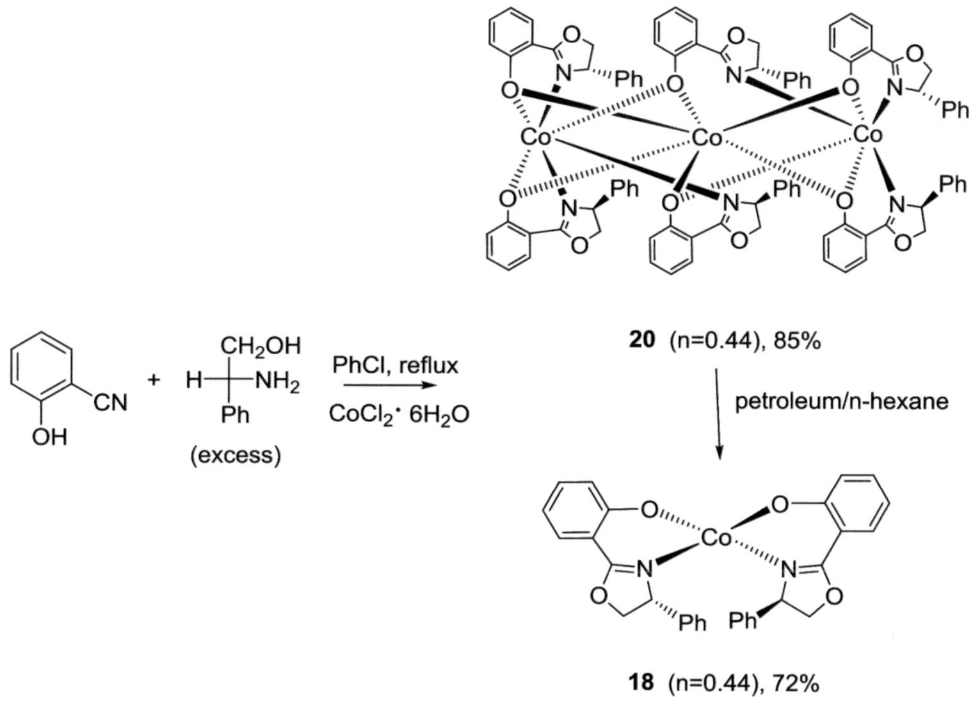

Figure 23.
*Solvent effects on the formation of complexes **18** and **20**.*

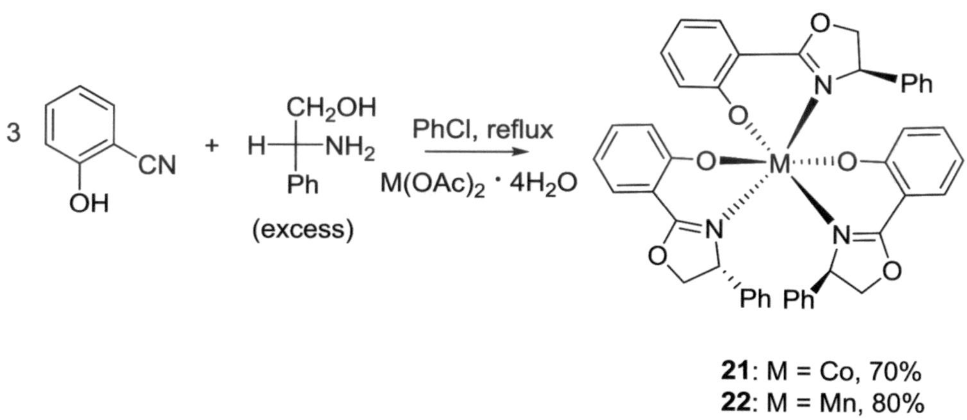

Figure 24.
*One-pot synthesis of tri(oxazoline) metal complexes **21** and **22**.*

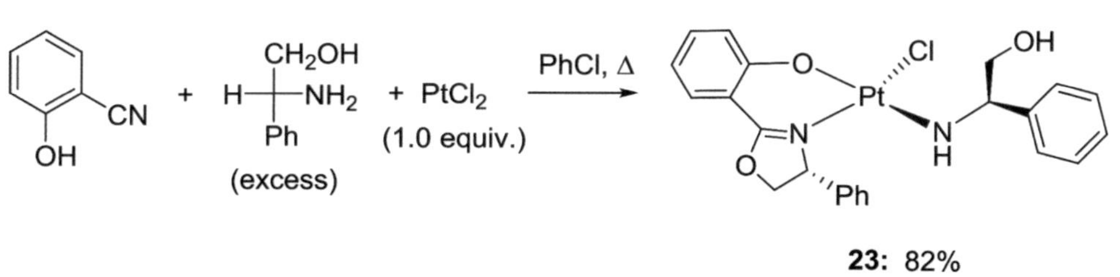

Figure 25.
*One-pot synthesis of oxazoline platinum complex **23**.*

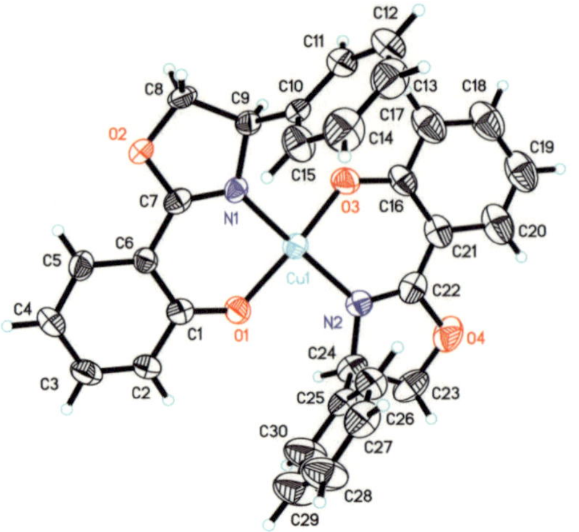

Figure 26.
The crystal structure of complex 16.

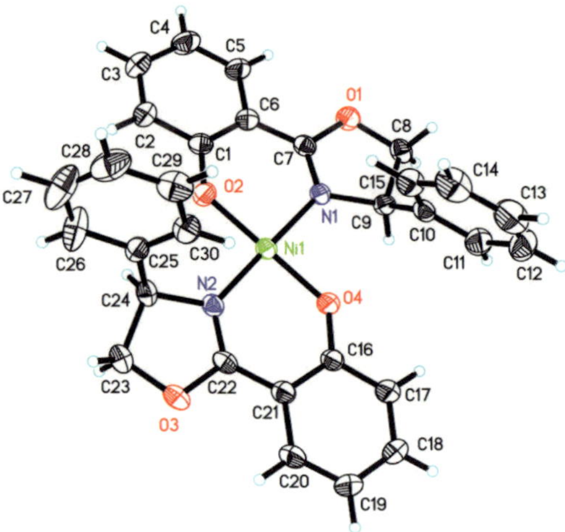

Figure 27.
The crystal structure of complex 17.

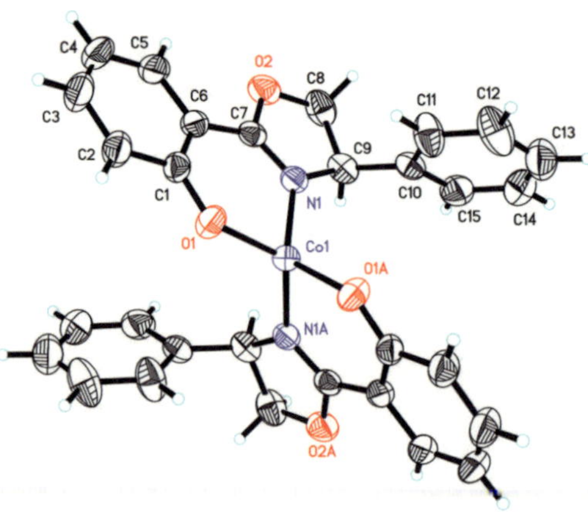

Figure 28.
The crystal structure of complex 18.

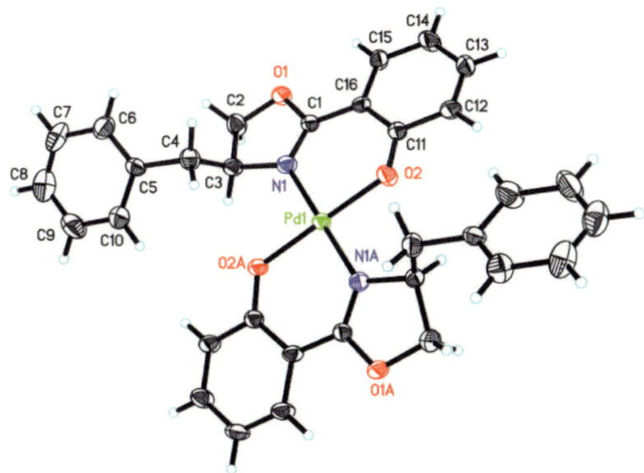

Figure 29.
The crystal structure of complex 19.

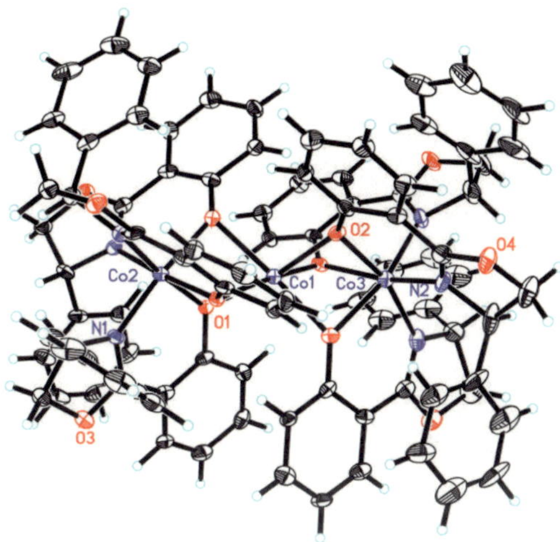

Figure 30.
The crystal structure of complex 20.

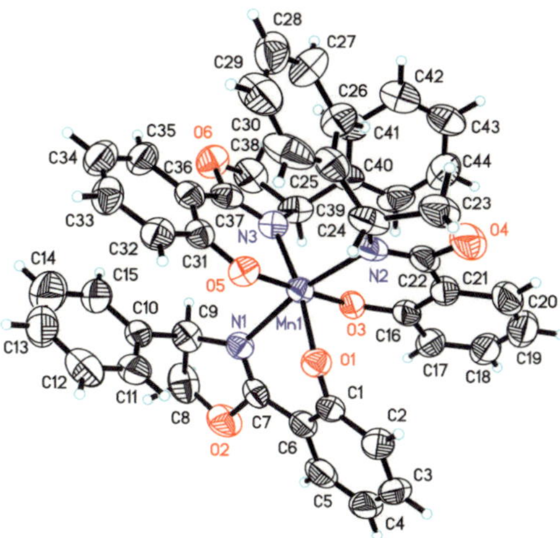

Figure 31.
The crystal structure of complex 21.

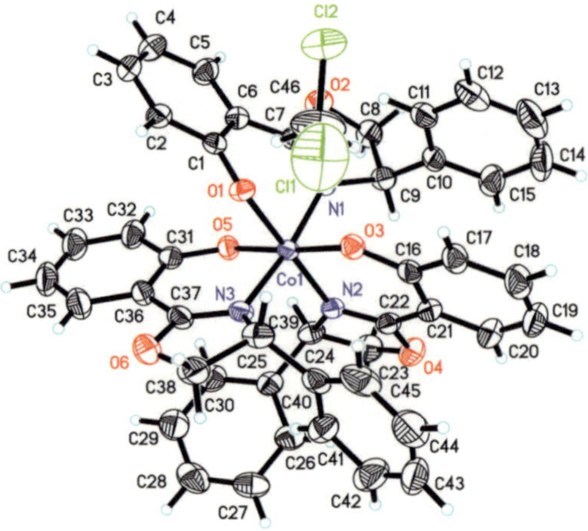

Figure 32.
The crystal structure of complex 22.

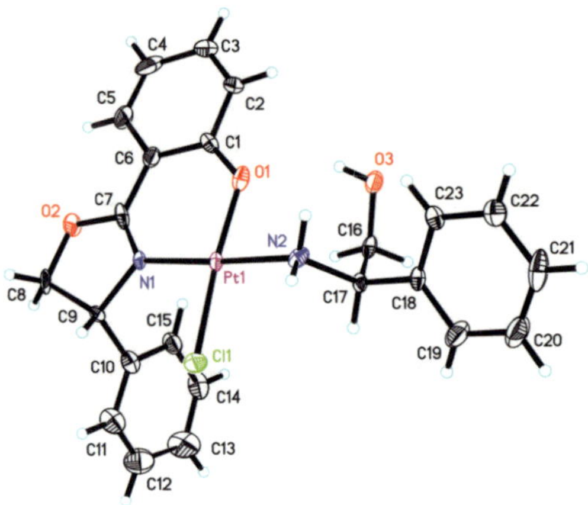

Figure 33.
The crystal structure of complex 23.

Metal salts	The amount of metal salts	Products (%)	Yield (%)
$Cu(OAc)_2{\cdot}H_2O$	55.7	16	65
$CuCl_2{\cdot}2H_2O$	53.2	16	85
$Ni(OAc)_2{\cdot}4H_2O$	51.0	17	92
$NiCl_2{\cdot}6H_2O$	53.0	17	95
$CoCl_2{\cdot}6H_2O$	44.3	18, 20	72, 85
$PdCl_2$	49.8	19	86
$Co(OAc)_2{\cdot}4H_2O$	42.3	21	70
$Mn(OAc)_2{\cdot}4H_2O$	52.6	22	80
$PtCl_2$	33.7	23	82

Table 2.
One-pot synthesis of salicyloxazoline complexes.

one unit of (R)-2-(4-phenyl-4,5-dihydrooxazol-2-yl)phenol and one unit of D-phenylglycinol was obtained after column chromatography with petroleum ether and dichloromethane (4:1) followed by crystallization via slow evaporation (**Figures 25** and **33**).

The proposed mechanism shows that excess metal salts can activate the reaction of 2-hydroxybenzonitrile with D-phenylglycinol in chlorobenzene to form ligand intermediates and then directly furnish the corresponding organometallic complexes via a one-step procedure (**Table 2**).

4. Conclusions

One-pot synthesis of oxazolinyl-zinc(II) complexes **1–23** at yields 25–95% was firstly demonstrated by assembling three-component reactions between metal salts, amino alcohols, and a variety of nitrile precursors. From the crystal structures of the complexes **1–23**, the reaction product is highly dependent on the presence of ligands, the amount of metal salts, and the nature of the substituent at the stereogenic center, giving a variety of coordination modes, such as mono- and bis-chelate complexes, mononuclear and mutinuclear complexes, etc.

Investigations into other oxazolinyl organometallic complexes and the catalytic properties of these complexes as chiral ligands are currently ongoing. These complexes exhibit bioactivities as anticancer reagents, and the future use of these complexes in medical fields is currently being developed.

Acknowledgements

This work was supported by the Hefei University of Technology and the University of Science and Technology of China. The authors also thank prof. King Kuok (Mimi) from Imperial College London, prof. Peter J. Stang from the University of Utah and prof. K.H. Lee from the University of Carolina at Chapel Hill for providing supports and help.

Conflicts of interest

The authors declare that they have no competing interests.

Appendix

Additional information is provided in the article as follows:

1. The characterization spectra of compounds 1–**15** [44] are available free of charge via the Internet at https://ccj.springeropen.com/. The crystallographic information of compounds **1**–**15** are available from the Cambridge Crystallographic Data Center (CCDC) as supplementary publications CCDC 853709–853,710, 931, 745–931,746, 931,745–931,748, 931,751–931,753, 931,756, 1,014,806–1,014,807, and 1,540,756, deposit@ccdc.cam.ac.uk or http://www. ccdc.cam.ac.uk.

2. Supporting information including the NMR spectra for compounds **1–15** [44] are available free of charge via the Internet at https://ccj.springeropen.com/articles/10.1186/s13065-017-0305-1.

3. Supporting information including the NMR spectra for compounds **16–23** [45] are available free of charge via the Internet at: Crystallographic Data Center (CCDC) as supplementary publications CCDC 853709–853710, 931745–931746, 931745–931748, 931751–931753, 931756, 1014806–1014807 and 1540756.

4. The crystallographic information of compounds **1–23** are available from the Cambridge Crystallographic Data Center (CCDC) as supplementary publications CCDC 853709–853,710, 931,745–931,746, 931,745–931,748, 931,751–931,753, 931,756, 1,014,806–1,014,807, 1,540,756, and CCDC 1035470–1,035,473, 1,035,475–1,035,476, 1,035,626, and 1,835,179, deposit@ccdc.cam.ac.uk or http://www.ccdc.cam.ac.uk.

Author details

Mei Luo
Hefei University of Technology, Hefei, Anhui, China

*Address all correspondence to: luomei@pu.edu.cn

References

[1] Desimoni G, Faita G, Jorgensen KA. Update 1 of C-2-symmetric chiral bis(oxazoline) ligands in asymmetric catalysis. Chemical Reviews. 2011;**111**:284-437. DOI: 10.1021/cr100161x

[2] Hargaden GC, Guiry PJ. Recent applications of oxazoline-containing ligands in asymmetric catalysis. Chemical Reviews. 2009;**109**:2505-2550. DOI: 10.1021/cr800400z

[3] O'Reilly S, Guiry PJ. Recent applications of C-1-symmetric bis(oxazoline)-containing ligands in asymmetric catalysis. Synthesis-Stuttgart. 2014;**46**:722-739. DOI: 10.1055/s-0033-1340829

[4] Gotoh R, Yamanaka M. Chiral Zn(II)-bisamidine complex as a Lewis-Bronsted combined acid catalyst: Application to asymmetric Mukaiyama aldol reactions of alpha-ketoesters. Molecules. 2012;**17**:9010-9022. DOI: 10.3390/molecules17089010

[5] Makino K, Ogawa I, Hamada Y. Synthesis of new chiral bis-oxazoline ligand with zinc triflate-selective chelating ability and its applications. Heterocycles. 2005;**66**:433-440. DOI: 10.3987/COM-05-S(K)49

[6] Abbina S, Du G. Zinc-catalyzed highly isoselective ring opening polymerization of rac-lactide. ACS Macro Letters. 2014;**3**:689-692. DOI: 10.1021/mz5002959

[7] Abbina S, Du G. Chiral amido-oxazolinate zinc complexes for asymmetric alternating copolymerization of CO_2 and cyclohexene oxide. Organometallics. 2012;**31**:7394-7403. DOI: 10.1021/om3006992

[8] Bleith T, Deng Q-H, Wadepohl H, Gade LH. Radical changes in Lewis acid catalysis: Matching metal and substrate. Angewandte Chemie, International Edition. 2016;**55**:7852-7856. DOI: 10.1002/anie.201603072

[9] Le Roux E, Merle N, Tornroos KW. Synthesis and characterisation of trigonal C-2-chiral di- and tetra-substituted bis(oxazoline) alkylzinc complexes and their reactivity towards protic reagents. Dalton Transactions. 2011;**40**:1768-1777. DOI: 10.1039/C0DT01252A

[10] Witte H, Seeliger W. Formation of cyclic imidic esters by reaction of nitriles with amino alcohols. Liebigs Annalen der Chemie. 1974:996-1009. DOI: 10.1002/jlac.197419740615

[11] Bolm C, Weickhardt K, Zehnder M, Ranff T. Synthesis of optically active bis(2-oxazolines): Crystal structure of a 1,2-bis(2-oxazolinyl)benzene $ZnCl_2$ complex. Chemische Berichte. 1991;**124**:1173-1180. DOI: 10.1002/cber.19911240532

[12] Kögel JF, Kusaka S, Sakamoto R, Iwashima T, Tsuchiya M, Toyoda R, et al. Heteroleptic [bis(oxazoline)](dipyrrinato)zinc(ii) complexes: Bright and circularly polarized luminescence from an originally achiral dipyrrinato ligand. Angewandte Chemie, International Edition. 2016;**55**:1377-1381. DOI: 10.1002/anie.201603072

[13] Telfer SG, McLean TM, Waterland MR. Exciton coupling in coordination compounds. Dalton Transactions. 2011;**40**:3097-3108. DOI: 10.1039/C0DT01226B

[14] Rechavi D, Lemaire M. Enantioselective catalysis using heterogeneous bis(oxazoline) ligands: Which factors influence the enantioselectivity? Chemical Reviews. 2002;**102**:3467-3494. DOI: 10.1021/cr020008m

[15] Johnson JS, Evans DA. Enantioselective catalysis using heterogeneous bis(oxazoline) ligands: Which factors influence the enantioselectivity? Accounts of Chemical Research. 2000;**33**:325-335. DOI: 10.1021/ar960062n

[16] Pfaltz A. Chiral semicorrins and related nitrogen heterocycles as ligands in asymmetric catalysis. Accounts of Chemical Research. 1993;**26**:339-345. DOI: 10.1021/ar00030a007

[17] Ghosh AK, Mathivanan P, Cappiello J. C2-symmetric chiral bis(oxazoline)–metal complexes in catalytic asymmetric synthesis. Tetrahedron: Asymmetry. 1998;**9**:1-45. DOI: 10.1016/S0957-4166(97)00593-4

[18] Bolm C, Luong TKK, Schlingloft G. Enantioselective metal-catalyzed Baeyer-Villiger oxidation of cyclobutanones. Synlett. 1997:**10**;1151-1152

[19] Bolm C, Bienewald F, Schlingloft G. Copper- and vanadium-catalyzed asymmetric oxidations. Journal of Molecular Catalysis A: Chemical. 1997;**117**:347-350. DOI: 10.1016/S1381-1169(96)00359-7

[20] Bolm C, Bienewald F, Harms K. Syntheses and vanadium complex of salen-like bissulfoximines. Synlett. 1996;**8**:775-776. DOI: 10.1055/s-1996-5512

[21] Bolm C, Bienewald F. Asymmetric sulfide oxidation with vanadium catalysts and H_2O_2. Angewandte Chemie, International Edition in English. 1995;**34**:2640-2642. DOI: 10.1002/ange.19951072317

[22] Moreno RM, Bueno A, Moyano A. 4-Ferrocenyl-1,3-oxazoline derivatives as ligands for catalytic asymmetric allylation reactions. Journal of Organometallic Chemistry. 2003;**671**:187-188. DOI: 10.1016/S0022-328X(03)00095-0

[23] Braunstein P, Naud F. Hemilability of hybrid ligands and the coordination chemistry of oxazoline-based systems. Angewandte Chemie, International Edition. 2001;**40**:680-699. DOI: 10.1002/1521-3773(20010216)40:4<680::AIDANIE6800>3.0.CO;2-0

[24] Bolm C. Bis(4,5-dihydrooxazolyl) derivatives in asymmetric catalysis. Angewandte Chemie, International Edition in English. 1991;**30**:542-543. DOI: 10.1002/anie.199105421

[25] Gómez M, Muller G, Rocamora M. Coordination chemistry of oxazoline ligands. Coordination Chemistry Reviews. 1999;**193-195**:769-835. DOI: 10.1016/S0010-8545(99)00086-7

[26] Cozzi PG, Floriani C, Chiesi-Villa A, Rizzoli C. Oxazoline early transition metal complexes: Functionalizable achiral titanium(IV), titanium(III), zirconium(IV), vanadium(III), and chiral zirconium(IV) bis(oxazoline) complexes. Inorganic Chemistry. 1995;**34**:2921-2930. DOI: 10.1021/ic00115a020

[27] Cozzi PG, Gallo E, Floriani C, Chiesi-Villa A, Rizzoli C. (Hydroxyphenyl)oxazoline: A novel and remarkably facile entry into the area of chiral cationic alkylzirconium complexes which serve as polymerization catalysts. Organometallics. 1995;**14**:4994-4996. DOI: 10.1021/om00011a010

[28] Gant TG, Meyers AI. The chemistry of 2-oxazolines (1985–present), (hydroxyphenyl)oxazoline: A novel and remarkably facile entry into the area of chiral cationic alkylzirconium complexes which serve as polymerization catalysts. Tetrahedron. 1994;**50**:2297-2360. DOI: 10.1016/S0040-4020(01)86953-2

[29] Bolm C, Schlingloff G. Metal-catalyzed enantiospecific aerobic oxidation of cyclobutanones. Chemical Communications. 1995:**26**;1247-1248

[30] Bolm C, Schlingloff G, Weickhardt K. Optically-active lactones from a Baeyer-Villiger-type metal-catalyzed oxidation with molecular-oxygen. Angewandte Chemie, International Edition in English. 1994;**33**:1848-1849. DOI: 10.1002/anie.199418481

[31] Peng Y, Feng X, Yu K, Li Z, Jiang Y, Yeung C-H. Synthesis and crystal structure of bis-[(4S,5S)-4,5-dihydro-4,5-diphenyl-2-(2′-oxidophenyl-chi O)oxazole-chi N] copper(II) and its application in the asymmetric Baeyer-Villiger reaction. Journal of Organometallic Chemistry. 2001;**619**:204. DOI: 10.1016/S0022-328X(00)00687-2

[32] Bolm C, Luong TK, Harms K. Bis[2-(oxazolinyl)phenolato]oxovanadium(IV) complexes: Syntheses, crystal structures and catalyses. Chemische Berichte/Recueil. 1997;**130**:887-890. DOI: 10.1002/cber.19971300712

[33] Hwang JJH, Abu-Omar MM. New vanadium oxazoline catalysts for epoxidation of allylic alcohols. Tetrahedron Letters. 1999;**40**:8313-8316. DOI: 10.1016/S0040-4039(99)01711-6

[34] Gómez M, Jansat S, Muller G, Noguera G, Teruel H, Moliner V, et al. First dioxomolybdenum(VI) complexes containing chiral oxazoline ligands: Synthesis, characterization and catalytic activity. European Journal of Inorganic Chemistry. 2001:**4**;1071-1076

[35] Arias J, Newlands CR, Abu-Omar MM. Kinetics and mechanisms of catalytic oxygen atom transfer with oxorhenium(V) oxazoline complexes. Inorganic Chemistry. 2001;**40**:2185-2192. DOI: 10.1021/ic000917z

[36] Hoogenraad M, Kooijman H, Spek AL, Bouwman E, Haasnoot JG, Reedijk J. Manganese oxidation catalysts with 2-(2′-hydroxyphenyl)oxazoline ligands: Catalyst intermediates and degradation products—Crystal structure of bis[{2-(2′-oxazolinyl)phenolato}bis(1-methylimidazole)manganese(iii)] perchlorate. European Journal of Inorganic Chemistry. 2002:**11**;2897-2903

[37] Hoogenraad M, Ramkisoensing K, Gorter S, Driessen WL, Bouwman E, Haasnoot JG, et al. Electrochemical and catalytic properties of novel manganese(III) complexes with substituted 2-(2′-hydroxyphenyl)oxazoline ligands—X-ray structures of tris[5-methyl-2-(2′-oxazolinyl)phenolato]manganese(III) and tris[5-chloro-2-(2′-oxazolinyl)phenolato]manganese(III). European Journal of Inorganic Chemistry. 2002:**2**;377-387

[38] Hoogenraad M, Ramkisoensing K, Driessen WL, Kooijman H, Spek AL, Bouwman E, et al. Catalytic and electrochemical properties of new manganese(III) compounds of 2-(2′-hydroxyphenyl)-oxazoline (Hphox or HClphox). Molecular structures of $[Mn(Clphox)_2(MeOH)_2](ClO_4)$ and $[Mn(phox)_2(MeOH)_2][Mn(phox)_2(ClO_4)_2](H_2O)_2$. Inorganica Chimica Acta. 2001;**320**:117-126. DOI: 10.1016/S0020-1693(01)00485-6

[39] Hoogenraad M, Ramkisoensing K, Kooijman H, Spek AL, Bouwman E, Haasnoot JG, et al. Catalytic and electrochemical properties of new manganese(III) compounds of 2-(2′-hydroxyphenyl)-oxazoline (Hphox or HClphox). Molecular structures of $[Mn(Clphox)_2(MeOH)_2](ClO_4)$ and $[Mn(phox)_2(MeOH)_2][Mn(phox)_2(ClO_4)_2]'(H_2O)_2$. Inorganica Chimica Acta. 1998;**27**:217-220. DOI: 10.1016/S0020-1693(98)00118-2

[40] Miller KJ, Baag JH, Abu-Omar MM. Synthesis, characterization, and reactivity of palladium(ii) salen and oxazoline complexes. Inorganic Chemistry. 1999;**38**:4510-4514. DOI: 10.1021/ic981450j

[41] Đaković S, Liščić-Tumir L, Kirin SI, Vinković V, Raza Z, Sÿuste A, et al. Enantioselectivity in cyclopropanation catalyzed by Cu(I) complexes increased by π stacking of two monodentate oxazoline ligands. Journal of Molecular Catalysis A: Chemical. 1997;**18**:27-31. DOI: 10.1016/S1381-1169(96)00113-6

[42] Abu-Omar MM, McPherson LD, Arias J, Béreau VM. Clean and efficient catalytic reduction of perchlorate. Angewandte Chemie, International Edition. 2000;**39**:4310-4332. DOI: 10.1002/1521-3773(20001201)39:23<4310::AID-ANIE4310>3.0.CO;2-D

[43] Hu Y, Khan MA, Nicholas KM. Chiral Pd-, Cu- and Ni-chelates and their utilities as catalysts in allylic acetoxylation of alkenes. Journal of Molecular Catalysis A: Chemical. 1994;**91**:319-334. DOI: 10.1016/0304-5102(94)00037-9

[44] Luo M, Zhang JC, Pang WM, Hii KK. One-step multicomponent synthesis of chiral oxazolinyl-zinc complexes. Chemistry Central Journal. 2017;**11**:81. DOI: 10.1186/s13065-017-0305-1

[45] Luo M, Zhang JC, Yin H, Wang CM, Morris-Natschke S, Lee K-H. One-step templated synthesis of chiral organometallic salicyloxazoline complexes. BMC Chemistry. 2019;**13**:51. DOI: 10.1186/s13065-019-0565-z

Section 3

Nucleation and Crystal Growth

Chapter 4

Nucleation

Wan Nur Athirah Mazli, Mohd Afnan Ahmad and Shafirah Samsuri

Abstract

Nucleation is one of the processes that involves at the beginning of the certain process like freezing, melting, boiling, condensation and crystallization. This process normally occurs in the industry, where it is involving the thermodynamic phase that involves work, energy and temperature. The nuclei growth happens when the initial phase changes to the other phase. Unfortunately, the detail for the theory of nucleation is not well-known around the people who are working in the industry, even though there are many reports or writings available. Thus, few types of nucleation like homogeneous and heterogeneous nucleation and the other theory of nucleation have been summarized in this chapter.

Keywords: homogeneous, heterogeneous, ice crystal

1. Introduction

Crystallization is the process where the molecules or tiny ice crystal are attached to each other to produce bigger formation of molecule or ice layer. Large cluster is stable compared to the smaller cluster as most of the molecules are far from the surface of the solution. It starts to form at the center of the cluster and slowly spread to the outward part of the cluster [1].

Besides, crystallization happens when there is phase change from the liquid to solid or crystalline form. This process happens when concentration of the solute already exceeds the equilibrium concentration or known as the supersaturated solution. The solution provides the driving force for the process and develops the growth of the particle but it not always develops the crystallization process because the supersaturated solution might be in metastable state. The process will happen when the limit of metastable has reached the limit and it depends on the kinetics system [2].

Nucleation is the process that happens in the crystallization process and it can be divided into few categories [2]. Ice nuclei appear at the solid surface when the fusion of heat is transferred by the conduction toward the cooled surface. By this time, the thin film starts to grow thicker until the ice crystal layer is formed. Most of the nuclei are developed in the high supersaturation and it attaches to form the ice crystal [3].

Activated process is known as nucleation because there is a barrier that the growing nucleus needs to be defeated [4]. The nucleation phenomenon is widely known in most aspects of science. Besides, it normally occurs at the beginning of the form of the ice crystal, snow, volcanic and the rainfall [5]. The nucleation process can be divided into a few parts as a different process has different conditions.

Other than that, it is important to understand more detail about the formation of ice for the nucleation and the supercooling condition in a science and technology field [6].

2. Primary nucleation

The supersaturated state achieved is the first step considered in the crystallization process. For supersaturation condition itself are not sufficient for a system to crystallize. The crystal needs to grow where new crystallization centers must exist in the solution. This formation of one phase in another, under conditions where a free energy barrier exists, is known as nucleation [1].

Nucleation is a random process which in two same situations or system nucleation will happen at a different time [4, 7]. Generally, the theory introduces in describing this phenomenon include its nature and behavior of the new thermodynamic phase is called classical nucleation theory (CNT). On the other hand, the experimental result of vapor to liquid nucleation cannot be described using CNT include the Argon by a few orders of magnitude for model substances [8].

For the formation of ice in the water below 0°C, if the system not changing with time and the nucleation occurs in one step. Thus, the probability of nucleation happens is denied through exponential decay as seen in radioactive decay. This scenario can be seen in the nucleation of ice in supercooled [9]. Another example is the self-organization process that forms like amyloid assembly associated with Alzheimer's diseases which the self-assemble system by the energy consumed such as microtubules in cells are referred to the nucleation and growth [10].

There are two types of nucleation namely the homogeneous or spontaneous nucleation and heterogeneous nucleation. This phenomenon happens when nuclei are formed perfectly in a clean solution where there are no any foreign particles. In any practical situation the presence of foreign particles includes impurity molecules, dust particles or ions refer to the heterogeneous nucleation. Both the nucleation process classified as primary nucleation [11].

Besides, primary nucleation refers to the early formation of a crystal in absence of any other crystal or it will not affect or influence the process if there any presence of crystal in a system that will happen in two conditions as stated above. In homogenous nucleation, the nucleation is not affected by solids and this include walls of crystallizer vessel and particles of any substance [12]. While the second one is the heterogeneous nucleation caused by the increasing rate of nucleation by the presence of other substance of solid particles would otherwise not be seen without the existence of these foreign particles [13].

For homogeneous nucleation, it is rarely happening because higher energy is needed to start the nucleation with the absence of solid particles. Primary nucleation has been modeled with the following power-law expression.

$$B = \frac{dN}{dt} = k_n(c - c^*)^n \tag{1}$$

Eq. (1) shows that the nucleation requires the rate constant, the instantaneous solute concentration, the concentration at the saturation phase and the empirical exponent which is generally ranges between 3 and 4.

Basically, primary nucleation theories can be used to study the metastable zone width (MSZW) data and kinetic of nucleation for the crystallization process. MSZW is explained as the gap of supersaturation between the supersolubility curve and solubility curve. The kinetics of primary nucleation of $NaNO_3$ for fractional crystallization process of high saline wastewater had been studied by Bian et al. MSZW of $NaNO_3$ was analyzed based on different stirring rate, present of seed crystals and cooling rate by using ultrasonic velocity sensor [14].

3. Secondary nucleation

The development of nuclei which refer to the influential of microscopic crystals in the magma is the phenomenon of secondary nucleation [15]. This nucleation happens when the crystal growth is introduced with a contact of other existing crystal or "seeds" [16]. For example, under a certain condition fluid shear forces are sufficient to produce secondary nuclei from an existing crystal surface [17]. Fluid shear nucleation happens when the fluid passes through the crystal with higher speed, sweep away the nuclei that would otherwise be incorporated into a crystal. The scenario led to the formation of a new crystal by the "swept away" nuclei. This contact nucleation proves to be the most effective and common method in nucleation.

In addition, this secondary nucleation depends on supersaturation. As stated Strickland-Constable that arises since the starting size distribution of potential secondary nuclei is depend on supersaturation. However, the particle produces at critical size ranged. This secondary nucleation has been performed in the agitated system that many been found in industrial crystallizers [18]. Sometimes, for needle breeding appear at high supersaturation where the dendrites may be expanding, or the needle is growing from existing crystal. As a result, when the crystal breaks the new centers for crystal growth are formed and this also refers to collision nucleation where arises from contact between two growing crystals or between a crystal with another solid surface [19].

The second nucleation gives many advantages includes lower kinetic order and rate-proportional to supersaturation which allows easy control without unstable operation, the nucleation occurs at low supersaturation which the growth rate is optimal for good quality, lower energy needed where the crystal strike avoids the breaking existing crystal to form new crystal, and the quantitative fundamental have already been isolated and are being incorporated into practical [20]. The following model, although somewhat simplified, is often used to model secondary nucleation.

$$B = \frac{dN}{dt} = k_1 M_T^j (c - c^*)^b \tag{2}$$

where k_1 is a rate constant, MT is the suspension density, j is an empirical exponent that can range up to 1.5, and b is an empirical exponent that can range up to 5.

Xue et al. stated that secondary nucleation is one of the main steps in the crystallization industrial. Most of the researchers are focused on the mechanism and the kinetics of the secondary nucleation of the aluminum hydroxide crystallization. Besides, it is related to the growth of the crystal process through the surface nucleation mechanism. Other than that, secondary nucleation is known as the removal-limited and chemical reaction controlled [21].

4. Homogeneous nucleation

Homogeneous nucleation is known when the nucleation is occurring without any favorable nucleation sites. Normally, the nucleation process happens randomly as well as spontaneously. It requires supercooling or superheating for the medium and it happens in the inner of the uniform substance with more difficulty. There will be the formation of an interface at the boundaries of a new phase when the nucleus is formed. To start the nucleation process, the supercooled phase is needed where the liquids need to be cooled down to below the melting temperature which is the maximum temperature of homogeneous nucleation but in the same time, the liquid need to make sure that it above the freezing temperature (homogeneous nucleation temperature) [22].

With the absence of external insoluble substance, supercooled water starts to nucleate homogeneously due to the variation temperature, density and pressure. This happens because of the endless formation of the ice cluster [23]. For homogeneous nucleation, after certain period of time when the nucleation begins, the crystals or bubbles are growing and become bigger and at the same time new crystal or bubble starts to nucleate. This can be shown through **Figure 1**.

According to the Abyzov et al., they approach new way to predict the homogeneous nucleation rates of silicate glass by estimate the nucleation rates. This is because the temperature of glass transition is closed to the maximum homogeneous rate. By using this approach, they can avoid some serious problem but in the same time they could not solved the difference between the experimental nucleation rate and the Classical Nucleation Theory, CNT [24].

Besides, Gibbs free energy is needed to determine the critical radius for homogeneous nucleation and it can be shown in Eq. (3). To find out the critical radius and the critical free energy, the formula of the Gibbs free energy need to differentiate [25]. Eqs. (4) and (5) show the critical radius and critical free energy.

$$\Delta G_{homo} = -\frac{4}{3}\pi r^3 \Delta G_V + 4\pi r^2 \gamma_{SL} \tag{3}$$

$$r^* = \frac{2\gamma_{SL}}{\Delta G_V} = \left(\frac{2\gamma_{SL} T_m}{L_V}\right)\frac{1}{\Delta T} \tag{4}$$

$$\Delta G^* = \frac{16\pi\gamma_{SL}^3}{3\Delta G_V^2} = \left(\frac{16\pi\gamma_{SL}^3 T_m^2}{3L_V^2}\right)\frac{1}{(\Delta T)^2} \tag{5}$$

The system can lower the free energy by the dissolution of the solid when the value of r is less than by r^*. Embryos are the unstable solid particle while nuclei are the stable solid particles with the value r is more than the value of r^* Besides, the nucleation rate for homogeneous nucleation also important as it will determine the time for nuclei to transfer into the nucleus. To calculate the nucleation rate, the population of critical embryos is needed and it can be shown in Eqs. (6) and (7) shows the formula of the nucleation rate for homogeneous nucleation. ΔG_r is the

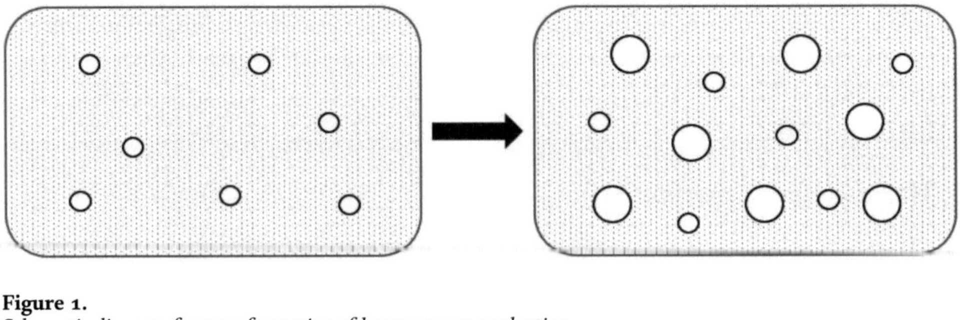

Figure 1.
Schematic diagram for transformation of homogeneous nucleation.

excess of free energy, k is the Boltzmann factor and n_0 is the total number of atoms in the system [25].

$$n_r = -\frac{4}{3}\pi r^3 \Delta G_v + 4\pi r^2 \gamma_{SL} \tag{6}$$

$$N_{homo} = f_0 C_0 e^{\frac{\Delta G^*_{homo}}{kT}} \tag{7}$$

5. Heterogeneous nucleation

Heterogeneous nucleation happens more often compared to the homogeneous nucleation. It is applicable for the phase transformation between the two-phase like bubble formation, solidification and the condensation of the gas. Normally, this process happens at a favorable site such as surfaces of the container, impurities and phase boundaries. At these sites, the effectiveness of the surface energy is lower; therefore, it can ease the nucleation process and reduce the free energy barrier.

The example of the heterogeneous nucleation is when the person is putting the finger into the carbonated water and the bubble is formed, this is because the surface of the finger is rougher compared to the surface of the container [26]. Besides, it happens when the ice nuclei at the lower temperature of supercooling [23]. For heterogeneous nucleation, after certain time when the nucleation starts to begin, the crystals or bubble only growing and become bigger and no new nucleation happens. **Figure 2** shows the transformation for heterogeneous nucleation.

As stated by Fujinaga and Shibuta, athermal heterogeneous nucleation is the process where the onset of the free growth of the phase is controlling the grain initiation of the inoculated particles. The onset of free growth is dependent on the particle sizes and undercooling temperature. In molecular dynamic simulation MD, additional condition should be chosen carefully in order to fix the heterogeneous [27].

For the heterogeneous nucleation process, the wetting will encourage the nucleation at the surfaces, thus the free energy is the same with the product of free energy for homogeneous and the contact angle (θ). Eq. (8) shows the free energy of the heterogeneous nucleation. It also has the critical radius and critical free energy and both of it can be shown through Eqs. (9) and (10).

$$\Delta G_{hete} = \Delta G_{homo} S(\theta) \tag{8}$$

$$r^* = \frac{2\gamma_{SL}}{\Delta G_V} \tag{9}$$

$$\Delta G^* = \frac{16\pi\gamma_{SL}^3}{3\Delta G_V^2} S(\theta) \tag{10}$$

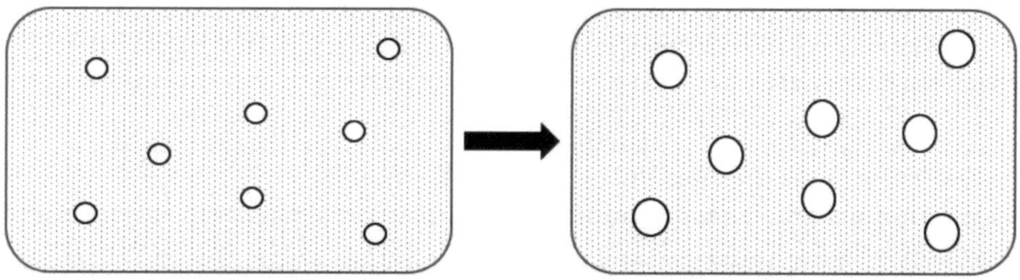

Figure 2.
Schematic diagram for transformation of heterogeneous nucleation.

To calculate the heterogeneous nucleation rate, it requires several atoms that contact with the surface of the nucleating agent surface. Eqs. (11) and (12) show the formula for the number of atoms and nucleation rates for heterogeneous nucleation. With minor undercooling situation, the nucleation process is happening in the cracks. To ensure that the cracks are effective, the opening of the crack should be large enough as it needs to let the solid to grow out freely without bothering the radius of the solid is smaller than the critical radius [25].

$$n^* = n_1 e^{-\frac{\Delta G_{hete}}{kT}} \tag{11}$$

$$N_{hete} = f_1 C_1 e^{-\frac{\Delta G^*_{hete}}{kT}} \tag{12}$$

6. Ice growth

The crystallization process for super-cooled water will start when the ice cluster solid has been formed by the nucleation. The ice starts to grow when the nucleation process has happened. Apart from it, the initial nucleus interrupts the second ice layer to growth when the latent heat is released [23].

Water forms the ices by undergoing the freezing process and this phase transition happens widely in nature. It is hard to redo the freezing process to transform water into solid with the specific crystalline structure [28]. The study of the growth of the ice layer on the subcooled surface started early in the field of geography. According to the Neumann problem, the phase change of water supposed to be uniform in the early process at the subcooled surface like ice crystal start to appear as a thin film around the surface. The ice starts thicken in the perpendicular direction uniformly. Primary nucleation will not happen simultaneously at the surface during the onset of the freezing on the subcooled surface [29].

The reaction of the ice particles in stratospheric clouds will change the halogen chemistry and at the same time, it helps to deplete the ozone layer. For the reactivity of the ice surface, it depends on its mobility or density as it is important to control the degree of the crystallinity and the structure of the ice film itself. To face this issue, ice film need to be grown well on the solid surfaces and use it as a tool to find out the reactions of the atmospheric. The structure of the ice film on the different hexagonal metal surface is sensitive to attach the energy to the metal and different structures of different transition metal surfaces [30].

For the phase change, heat and mass transfer is an important part of the onset like freezing on the subcooled surface. It is because it can be a good opportunity for the engineering application [29] such as ice formation on the cryogenic surface [31] and the ice production using scraped-surface heat exchanger [32]. For a certain process, the mass flow rate of the injected vapor is used to calculate the ice layer growth rate which is forms on the plate. Normally, the ice layer grows until it reaches the critical thickness is considered as the outcome of the process. Film-wise condensation starts to occur on the surface of the ice layer when it reaches the critical thickness [31].

The heat transfer coefficient in the scraped-surface heat exchanger without the phase change (ice growth) is three to five times smaller compared to the phase change. The torque for handle the scraper is increasing when the onset of ice growth is happening until the scrapper becomes frozen. Ice started to growth at stage II and stage III, where the ice particle from the nucleation takes place. The ice particle continuously freezes until it transforms into the ice crystal and the concentration of the mother liquor is increasing [33].

7. Kinetics of nucleation

As reviewed, the part of kinetics from the rate of nucleation gives a number of the water molecule which has a potential incorporated into the ice crystal [34]. The flux of water molecule that happens be incorporated with the ice crystal must be counted for nucleation rate estimation. The flux can be written as diffusive flux (Φ) and known as the Boltzmann distribution. The equation is expressed as below.

$$\Phi = \frac{K_B T}{h}.e^{\frac{\Delta g^{\#}}{K_B T}} \tag{13}$$

whereas h refers to the Planck's constant and $\Delta g^{\#}$ is the activation energy for the transfer of a water molecule across the water-ice boundary. The Φ can also be introduced as the frequency at water molecules overcome at the activation energy [34], as for the collisions rate or the probability in breaking the bond and reassemble of molecular.

Besides, the second pre-factor stated as the Zeldovich factor (Z) is for the depletion of the cluster population caused by crystal production. This factor includes the non-equilibrium of the kinetics process which the value is between 10^{-2} and 1, where the value 1 considers as assumed equilibrium. As a result, in describing the expression for the Φ and the pre-factor Z the rate (K) for water molecule is transformed into ice crystal can be expressed as equation below;

$$K = n_s.4\pi r^2_{crystal}.Z.\Phi \tag{14}$$

whereas n_s refer to the number of molecules in jumping the crystal surface of the neighboring crystal and the $4\pi r^2_{crystal}$ is the surface are of the critical ice crystal.

8. Conclusion

Most of the basic theories of the nucleation have been summarized in this chapter. The objective of this chapter is to introduce more about the primary, secondary, homogeneous, heterogeneous, ice growth and the kinetics of the nucleation. Even though the homogeneous and heterogeneous nucleation can be called as part of the primary nucleation, both of it has different characteristic as both processes happen at different situation and time. Basically, homogeneous always nucleate new crystals while heterogeneous only focus on the growth of the crystals. Both nucleations can be analyzed through the kinetics of the nucleation and the ice growth of the crystals or bubbles.

Acknowledgements

The authors would like to acknowledge Centre for Biofuel and Biochemical Research (CBBR) and also Chemical Engineering Department, Universiti Teknologi PETRONAS for the support.

Conflict of interest

The authors declare no conflict of interest.

Author details

Wan Nur Athirah Mazli[1,2], Mohd Afnan Ahmad[1,2] and Shafirah Samsuri[1,2*]

1 Chemical Engineering Department, Universiti Teknologi PETRONAS, Seri Iskandar, Perak, Malaysia

2 Centre for Biofuel and Biochemical Research, Universiti Teknologi PETRONAS, Seri Iskandar, Perak, Malaysia

*Address all correspondence to: shafirah.samsuri@utp.edu.my

References

[1] Lutsko JF. How crystals form: A theory of nucleation pathways. Science Advances. 2019;**5**(4):eaav7399

[2] Randall DG, Nathoo J. A succinct review of the treatment of reverse osmosis brines using freeze crystallization. Journal of Water Process Engineering. 2015;**8**:186-194

[3] Amran NA, Samsuri S, Ngadi N, Zakaria YZ, Jusoh M. Behaviour of ice crystal growth in a vertical finned cylindrical freeze concentrator. Applied Mechanics and Materials. 2014;**695**: 451-454

[4] Sear RP. Nucleation: Theory and applications to protein solutions and colloidal suspensions. Journal of Physics. Condensed Matter. 2007;**19**(3):033101

[5] Karthika S, Radhakrishnan TK, Kalaichelvi P. A review of classical and nonclassical nucleation theories. Crystal Growth & Design. 2016;**16**(11): 6663-6681

[6] Stan CA et al. A microfluidic apparatus for the study of ice nucleation in supercooled water drops. Lab on a Chip. 2009;**9**(16):2293-2305

[7] Pruppacher HR, Klett JD, Wang PK. Microphysics of clouds and precipitation. Atmospheric and Oceanographic Sciences Library. 1997; **18**(58–73):433-446

[8] Fladerer A, Strey S. Homogeneous nucleation and droplet growth in supersaturated argon vapor: The cryogenic nucleation pulse chamber. The Journal of Chemical Physics. 2006; **124**(16):164710

[9] Duft D, Leisner T. Laboratory evidence for volume-dominated nucleation of ice in supercooled water microdroplets. Atmospheric Chemistry and Physics. 2004;**4**(7):1997-2000

[10] Gillam JE, Macphee CE. Modelling amyloid fibril formation kinetics: Mechanisms of nucleation and growth. Journal of Physics. Condensed Matter. 2013;**25**(37):373101

[11] Geidobler R, Winter G. Controlled ice nucleation in the field of freeze-drying: Fundamentals and technology review. European Journal of Pharmaceutics and Biopharmaceutics. 2013;**85**(2):214-222

[12] Mitchell DL, Garnier A, Avery MA, Erfani E. CALIPSO observations of the dependence of homo- and heterogeneous ice nucleation in cirrus clouds on latitude, season and surface condition. Atmospheric Chemistry and Physics Discussions. 2016;**2016**:1-60

[13] Bi Y, Cao B, Li T. Enhanced heterogeneous ice nucleation by special surface geometry. Nature Communications. 2017;**8**:1-7

[14] Bian C, Chen H, Song X, Yu J. Metastable zone width and the primary nucleation kinetics for cooling crystallization of $NaNO_3$ from NaCl-$NaNO_3$-H_2O system. Journal of Crystal Growth. 2019;**518**:5-13

[15] Chianese A, Di Berardino F, Jones AG. On the effect of secondary nucleation on the crystal size distribution from a seeded batch crystallizer. Chemical Engineering Science. 1993;**48**(3):551-560

[16] Bergfors T. Seeds to crystals. Journal of Structural Biology. 2003; **142**(1):66-76

[17] Yousuf M, Frawley PJ. Experimental evaluation of fluid shear stress impact on secondary nucleation in a solution crystallization of paracetamol. Crystal Growth & Design. 2018;**18**(11): 6843-6852

[18] Ottens EPK, De Jong EJ. A model for secondary nucleation in a stirred vessel cooling crystallizer. Industrial and Engineering Chemistry Fundamentals. 1973;**12**(2):179-184

[19] Lal DP, Mason REA, Strickland-Constable RF. Collision breeding of crystal nuclei. Journal of Crystal Growth. 1969;**5**(1):1-8

[20] Fersht AR. Nucleation mechanisms in protein folding. Current Opinion in Structural Biology. 1997;**7**(1):3-9

[21] Xue J et al. Secondary nucleation and growth kinetics of aluminum hydroxide crystallization from potassium aluminate solution. Journal of Crystal Growth. 2019;**507**:232-240

[22] Zang L. Lecture 10: Homogeneous nucleation. University of Utah [Online]. Available from: https://my.eng.utah.edu/~lzang/images/lecture-10.pdf

[23] Schulte K, Weigand B. On the analytical modelling of the initial ice growth in a supercooled liquid droplet. International Journal of Heat and Mass Transfer. 2018;**127**:1070-1081

[24] Abyzov AS, Fokin VM, Zanotto ED. Predicting homogeneous nucleation rates in silicate glass-formers. Journal of Non-Crystalline Solids. 2018;**500**:231-234

[25] Darwish MS. Nucleation and Growth. American University of Beirut [Online]. Available from: https://www.aub.edu.lb/msfea/research/Documents/CFD-Nucleation.pdf.

[26] Zang L. Lecture 12: Heterogeneous nucleation – A surface catalyzed process. University of Utah [Online]. Available from: https://my.eng.utah.edu/~lzang/images/lecture-12.pdf

[27] Fujinaga T, Shibuta Y. Molecular dynamics simulation of athermal heterogeneous nucleation of solidification. Computational Materials Science. 2019;**164**:74-81

[28] Matsumoto M, Saito S, Ohmine I. Molecular dynamics simulation of the ice nucleation and growth process leading to water freezing. Nature. 2002;**416**:409-413

[29] Qin FGF, Chen XD, Free K. Freezing on subcooled surfaces, phenomena, modeling and applications. International Journal of Heat and Mass Transfer. 2009;**52**(5–6):1245-1253

[30] Haq S, Harnett J, Hodgson A. Growth of thin crystalline ice films on Pt (1 1 1). Surface Science. 2002;**505**:171-182

[31] Marshall T, Girard C. Modeling of ice formation and condensation on a cryogenic. Fusion Engineering and Design. 2001;**54**:473-483

[32] Martínez DS, Solano JP, Illán F, Viedma A. Analysis of heat transfer phenomena during ice slurry production in scraped surface plate heat exchangers. International Journal of Refrigeration. 2014;**48**:221-232

[33] Qin F, Chen XD, Ramachandra S, Free K. Heat transfer and power consumption in a scraped-surface heat exchanger while freezing aqueous solutions. Separation and Purification Technology. 2006;**48**:150-158

[34] Ickes L, Welti A, Hoose C, Lohmann U. Classical nucleation theory of homogeneous freezing of water: Thermodynamic and kinetic parameters. Physical Chemistry Chemical Physics. 2015;**17**(8):5514-5537

Chapter 5

Synthesis Methods and Crystallization of MOFs

Yitong Han, Hong Yang and Xinwen Guo

Abstract

Metal-organic frameworks (MOFs) are a class of porous crystalline materials constructed of metal centres with organic linkers, creating one-, two-, or three-dimensional well-organized frameworks with very high surface areas. The study of MOFs has become one of the research hot spots in many fields, owing to the broad potential applications projected for these materials in various areas. It is well recognized that synthesis strategies dictate the structure and thus the properties and performance of the resulted MOFs. This chapter provides a comprehensive up-to-date overview on the modulated synthesis strategies for MOFs. The ability to control crystal morphology and size by a number of modulated synthesis methods is illustrated by the zirconium-terephthalate-based MOF, the UiO-66, and a number of other MOFs.

Keywords: metal-organic frameworks, synthesis method, crystallization, crystal morphology and size control

1. Introduction

Porous materials are a class of solid compounds with an ordered and/or disordered pore structure, high pore volume, and large surface area. These combined with some of their unique chemistries make them a unique class of chemical and engineering materials. Over the past three decades, porous materials have become one of the research hot spots in the fields of chemistry, physics, and materials science and engineering. The diversity in the pore orientation and dimensionality, combined with the multiplicity in pore shapes and sizes, makes the porous materials highly interested and widely studied. The IUPAC defines the porous materials based on the pore dimensions and classifies them into the microporous materials with pore dimension less than 2 nm, mesoporous materials with pore dimension ranging from 2 to 50 nm, and macroporous materials with pores larger than 50 nm [1].

Porous materials have shown great application values in some traditional industries, such as the oil and gas processing, industrial catalysis, adsorption/separation, and fine chemical industries [2–5]. In more recent decade, they are also being recognized and explored in the research fields of water treatment, sustained release of drugs, and fuel cells, to name a few [6, 7].

Among all porous materials, zeolites are perhaps the most famous microporous materials. In the frameworks of the zeolites, different numbers of TO_4 tetrahedral primary structure units connect together to form shared apexes, which enable the creation of the interconnected secondary structure units of various shapes and the formation of the microporous zeolite structure. The "T" in the TO_4 tetrahedrons is usually silicon, aluminum, and phosphorus. Chemically, zeolites have adjustable acidity, in addition to their excellent pore channel selectivity and hydrothermal stability. These support the high research value of zeolites in industrial applications such as exchange, adsorption/separation, and petroleum processing. However, the inorganic nature of zeolites limits their adjustability in chemical properties and designability in pore structures and sizes, which restrict the further development and applications of zeolites [2].

For decades, there have been research efforts trying to integrate various metal centres and functional organic molecules into porous structures for the modulation of physical and chemical properties. This effort has resulted in the development of a novel type of hybrid porous crystalline materials formed by the coordination of inorganic metal centres and organic linkers in the 1990s. This new class of porous materials is most commonly known as metal–organic frameworks (MOFs), although they are also sometimes known as porous coordination polymers (PCPs) or inorganic-organic hybrid materials [8–10]. Ever since their discovery, MOFs have received significant attention from scientists, engineers, and technologists due to their versatilities in structures and chemistries. Since the synthesis of MOFs directly influences the crystallization of the MOF structure, thus dictates its properties and functional performance, extensive work and thousands of research papers have continually emerged focusing on the development of synthesis methods.

The aim of this chapter is to provide an up-to-date overview on the synthesis methods of MOFs reported so far, with the objective to provide empirical guidance for developing synthesis strategy for MOF materials targeting specific properties and functionalities. We draw on our own extensive experiences on modulated synthesis of the UiO-66 MOF material to demonstrate how the crystal morphology and size can be controlled by different modulated synthesis methods, mechanisms of crystallization, and effect of metal doping on MOF crystals during synthesis.

2. Structure and applications of MOFs

As stated in the introduction, MOFs are a unique class of hybrid porous crystalline materials and have been widely studied over the past two decades for their inherent structure design flexibility and potential applications [11].

The structures of MOFs are constructed by self-assembly between the "nodes" of metal-containing secondary building units (SBUs) and the "bridges" of organic linkers, creating one-, two-, or three-dimensional well-organized network structures with very high pore volumes and surface areas. The framework topologies and pore structures and sizes of MOFs can be designed via selecting various metal centres and organic linkers. Their chemical properties can be modified by chemical functionalization of linkers and post modifications [12–14].

MOFs have extended the chemistry of the porous materials from inorganic to inorganic-organic hybrid. This compositional diversity, combined with the structural diversity, gives them unique properties and functionalities. MOFs are thus a class of highly attractive porous materials for a broad range of applications, including in gas adsorption/separation [15–18], luminescence and sensing [19–22], catalysis [23–25], and others [26–29].

3. Synthesis methods and crystallization of MOFs

3.1 Overview of synthesis methods for MOFs

At their discovery, the method for synthesis of MOFs is solvothermal. Typically, metal precursors and organic linkers are dissolved in solvent and placed in a closed reaction vessel for the formation and self-assembly of MOF crystals. The common solvents used include N,N-dimethylformamide (DMF), N,N-diethylformamide (DEF), methanol, ethanol, and acetonitrile. The synthesis temperature is generally below 220°C, and the crystallization time varies from several hours to several tens of days.

After more than two decades of research and development, great advances have been made in the synthesis of MOFs. New synthesis methods such as the electrochemical, microwave-assisted, mechanochemical synthesis, microfluidic synthesis method, etc. have all been reported [30]. **Figure 1** summarizes the development timeline for the most common synthesis approaches of MOFs [30].

These diverse methods have enabled the synthesis of hundreds of new MOF structures since its first discovery. On top of this, as clearly stated in the principle of "structure dictates function" [31], having the ability to control and tailor the morphology and size and their chemical functionalization of MOF crystals is vital in delivering targeted properties and performances of the resulted MOF materials. This demands the development of more sophisticated synthesis strategies based on the understanding of mechanisms of crystallization occurred during synthesis. The following provides an overview on the currently developed modulated synthesis methods for morphology and size control of MOF crystals and doping to create hybrid MOF crystals.

3.2 Morphology and size control of MOF crystals

The synthesis of MOFs involves the process of crystallization during which the nucleation and growth of crystals occur. The nucleation and growth of MOF crystals involve the self-assembly between metal-oxygen clusters and organic linkers. Understanding the influencing factors on the nucleation and growth of MOF crystals during their synthesis will enable accurate controlling of crystal morphology and size. The following discusses the morphology and size development of the MOF crystals during different modulated syntheses.

3.2.1 Deprotonation regulation synthesis

It is well documented that synthesis conditions, such as temperature, time, solvent type, and reactant concentrations, play important roles in the morphology and size of resulted MOF crystals.

For example, an NH_2-MIL-125(Ti) MOF material can be synthesized by a solvothermal method in a mixed solvent of DMF and methanol. **Figure 2** shows the SEM images of the NH_2-MIL-125(Ti) crystals synthesized with different reactant concentrations, as indicated by the total solvent volumes. By changing the total volume of the solvent alone while maintaining constant ratio between the DMF and methanol and amount of reactants, the morphology of the NH_2-MIL-125(Ti)

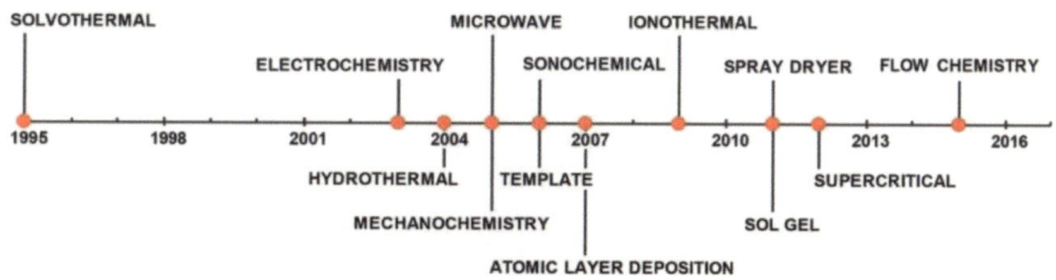

Figure 1.
Timeline of the most common synthesis approaches for MOFs [30].

crystals can be modified from circular plates to tetragons and octahedrons [32]. As the crystal morphology changes, the light absorption band of the NH_2-MIL-125(Ti) can be tuned from 480 to 533 nm, making it advantageous in photocatalytic applications.

It has been found that the reactant concentration has a significant effect on the deprotonation rate of the organic linkers during the synthesis of NH_2-MIL-125(Ti) crystals. The deprotonation rate plays a critical role in the nucleation and growth of the NH_2-MIL-125(Ti) crystals. Modulating crystal morphology and size of MOFs by changing the rate of deprotonation is called the deprotonation regulation synthesis [33].

3.2.2 Coordination modulation synthesis

Besides changing synthesis conditions, introducing additives during crystallization is another strategy of controlling the morphology and size of MOF crystals. The research lead by Kitagawa first reported the coordination modulation of carboxylic acid additives on the growth of $[\{Cu_2(ndc)_2(dabco)\}_n]$ MOF crystals [34]. In this work, a specific amount of acetic acid is added in the crystallization mother liquor of the $[\{Cu_2(ndc)_2(dabco)\}_n]$. Acetic acid hinders the coordination

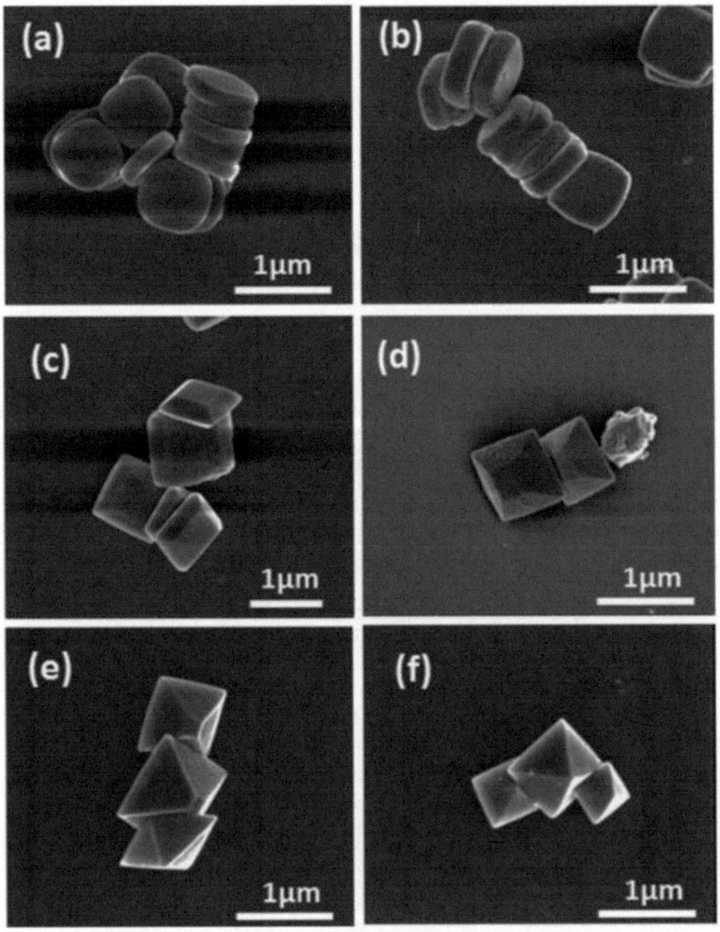

Figure 2.
SEM images of NH_2-MIL-125(Ti) crystals synthesized with different total solvent volumes of (a) 40 mL, (b) 30 mL, (c) 20 mL, (d) 15 mL, (e) 14 mL, and (f) 13.5 mL [32].

between metal clusters and organic linkers through selective coordination with the metal clusters. This influences the expansion of lattice structure and alters the growth of the $[\{Cu_2(ndc)_2(dabco)\}_n]$ crystals. **Figure 3** shows the nanocrystals of $[\{Cu_2(ndc)_2(dabco)\}_n]$ formed under the capping effect of acetic acid [34].

Subsequently, Kitagawa's group has also synthesized the HKUST-1 crystals using three kinds of monocarboxylic acid additives, the acetic acid, dodecanoic acid, and lauric acid under ultrasonic condition. **Figure 4** shows the effect of additive quantity on the morphology of the resulted HKUST-1 crystals. As seen, with the increased additive quantity, not only the size of the HKUST-1 crystals increases from tens of nanometres to several micrometres, the shape of the crystals also changes from cube to octahedron, truncated cube, and tetradecahedron, progressively [35, 36].

It has since been established that the coordination modulation which resulted from using the carboxylic acids has certain universality. It is shown to effectively control the crystal morphology and size of a number of MOFs, including $\{Cu_2(ndc)_2(dabco)\}_n$, HKUST-1, Zr-based MOFs, etc. The amount of the additive used ranges from 2 to 100 equimolar of the reactants. The carboxylic acid additives do not affect the crystal structure of the resulted MOFs.

3.2.3 Surfactant modulation synthesis

Some surfactants, such as the cetyltrimethylammonium bromide (CTAB) [37] and polyvinylpyrrolidone (PVDF) [38, 39], can also be used for the modulated synthesis of MOFs. In the crystallization of MOF crystals, the surfactant molecules can be selectively absorbed on one or more specific facets of the MOF crystals, thus hinder or alter their growth, and result in the modification of crystal morphology and size.

Take the hydrothermal synthesis of ZIF-67 MOF crystals as an example. As shown in **Figure 5**, by simply changing the amount of CTAB from 0.0025 to 0.025 wt%, a series of the ZIF-67 cubic or rhombic dodecahedron crystals with sizes ranging from ~150 nm to 1 μm can be produced [40]. Additionally, by carbonization of the resulted ZIF-67 crystals in N_2 flow, a series of Co-based porous carbon catalysts can be obtained. These Co-based porous carbon catalysts retain the

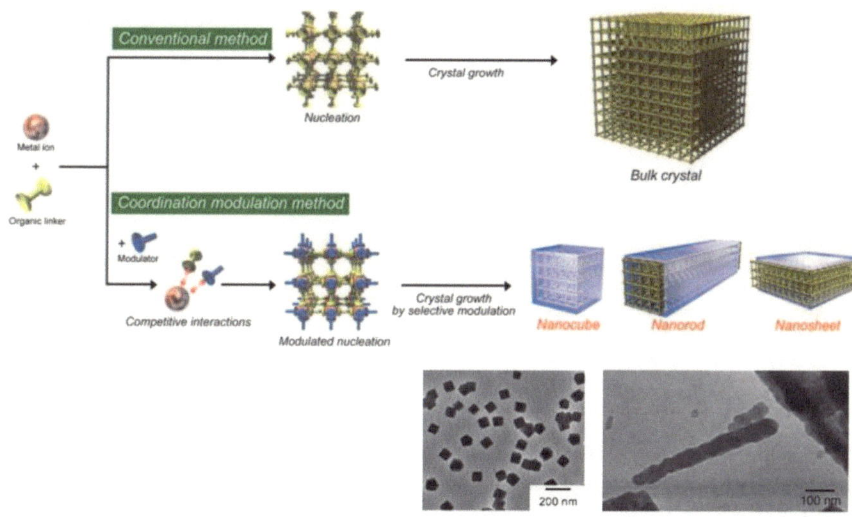

Figure 3.
Coordination modulation method for fabricating $[\{Cu_2(ndc)_2(dabco)\}n]$ nanocrystals [34].

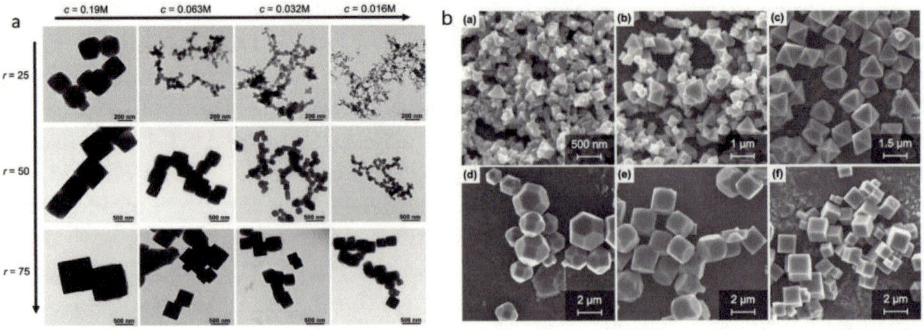

Figure 4.
(a) TEM images of the HKUST-1 crystals obtained with various concentrations of dodecanoic acid and H_3BTC; here C is the concentration of the H_3BTC and r is the molar ratio of dodecanoic acid to H_3BTC [35]. (b) SEM images of the HKUST-1 crystals obtained with different amounts of the lauric acid in mmol: (a) 0, (b) 2.34, (c) 4.75, (d) 7.13, (e) 9.5, and (f) 11.88 [36].

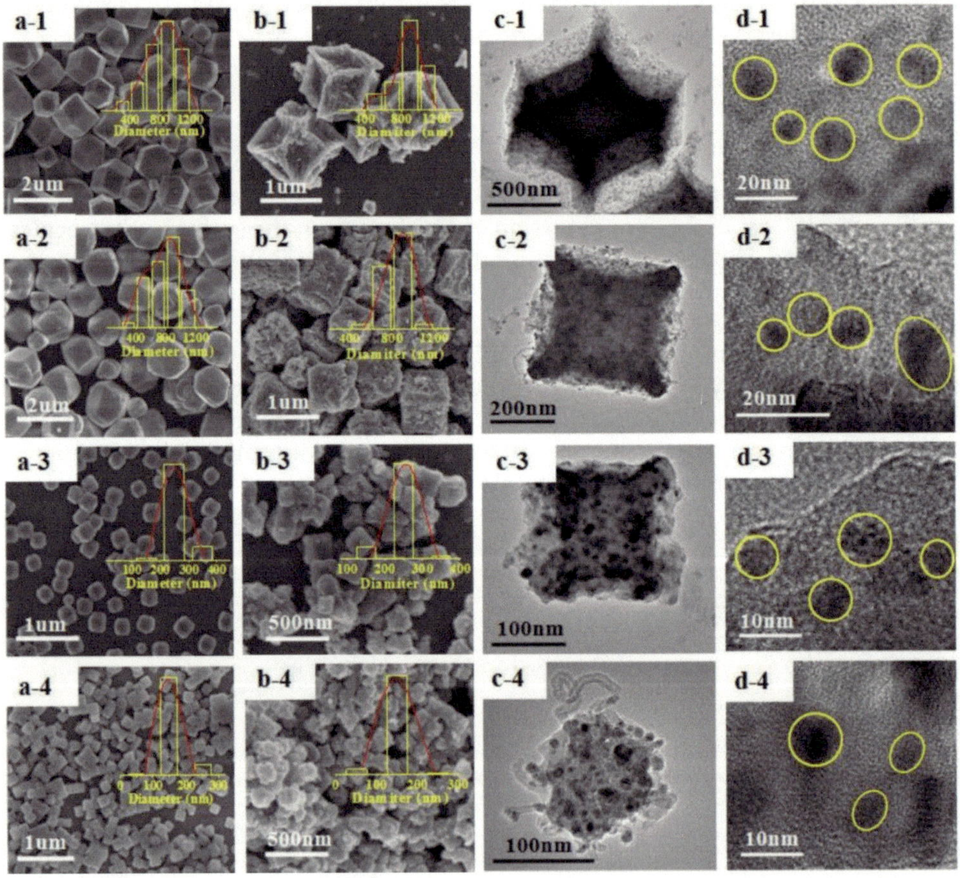

Figure 5.
ZIF-67 crystals synthesized with different amounts of the CTAB additive (1: 0, 2: 0.0025 wt%, 3: 0.01 wt%, and 4: 0.025 wt%). SEM and TEM images of (a) as-synthesized samples and (b–d) carbonized samples [40].

original shape of the ZIF-67 crystals and display an outstanding catalytic performance towards the CO_2 methanation at low temperatures.

It has also been reported that the addition of CTAB can modulate the crystal morphology and size of some other MOFs. **Figure 6** shows a number of such examples including IRMOFs [41, 42], HKUST-1 [43], and ZIF-8 [44].

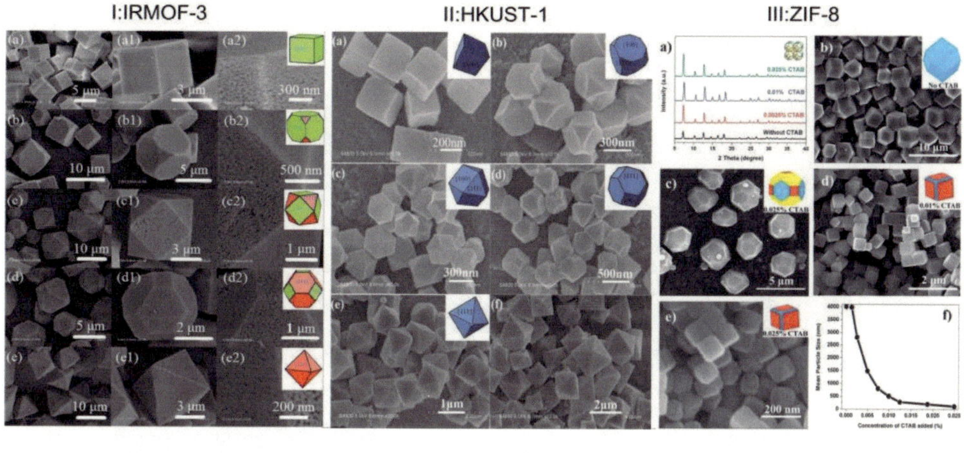

Figure 6.
The influence of CTAB additive on the crystal morphology and size of several MOFs. (I) SEM images of IRMOF-3 synthesized with different amounts of CTAB in mg: (a, a1, a2) 0, (b, b1, b2) 5, (c, c1, c2) 8, (d, d1, d2) 12, and (e, e1, e2) 15 [42]. (II) SEM images of HKUST-1 synthesized with different amounts of CTAB in M: (a) 0, (b) 0.005, (c) 0.01, (d) 0.05, (e) 0.1, and (f) 0.5 [43]. (III) (a) XRD patterns of ZIF-8; (b) ZIF-8 synthesized without CTAB; ZIF-8 synthesized with different amounts of CTAB in wt%: (c) 0.0025, (d) 0.010, and (e) 0.025; (f) mean particle size of ZIF-8 crystals versus the concentrations of CTAB added [44].

4. The UiO-66 MOF

As one of the most important MOFs, the Zr-BDC ($Zr_6O_4(OH)_4(CO_2)_{12}$) MOF, commonly known as the UiO-66, has been extensively studied due to its high porosity and excellent structural stability at high temperatures and pressures, and excellent stability in chemical (acidic/basic) aggressive environments. The framework of the UiO-66 is built on the [$Zr_6O_4(OH)_4$] SBUs, each coordinated with 12 1,4-benzene-dicarboxylate (H_2BDC) linkers. **Figure 7** illustrates the crystal structure of the UiO-66. Its cubic structure is composed of octahedral cages close to 11 Å and tetrahedral cages close to 8 Å, and these cages are connected through narrow triangular windows close to 6 Å [45].

4.1 Modulated synthesis of UiO-66 crystals using carboxylic acids

The UiO-66 crystals are commonly synthesized via a solvothermal method at 110–130°C and allowed to crystallize for 24 h. Without modulation, the obtained UiO-66 crystals are usually agglomerates of small cube-like particles of 80–200 nm in size and with low crystallinity. These UiO-66 crystals show low porosity with the BET surface area below 1000 $m^2\ g^{-1}$.

To improve the crystallinity of the UiO-66 crystals, carboxylic acid additives have been applied in the synthesis. Schaate et al. first studied the influence of benzoic acid and acetic acid on the crystal growth of the UiO-66 and other Zr-based MOFs [46]. They have found the UiO-66 crystals synthesized are octahedrons of several hundred nanometres in size, as shown in **Figure 8**. These crystals have a high crystallinity with the BET surface area of up to 1400 $m^2\ g^{-1}$. Schaate et al. suggested that the addition of the carboxylic acid additives changes the original coordination equilibrium and thus the crystal growth rate during the crystallization of the UiO-66 crystals. There exists a competition between the coordination of the BDC linkers and carboxylic acid additives towards the Zr_6 clusters. This becomes an obstacle for the connection of the BDC linkers and Zr_6 clusters, shifting the original coordination equilibrium. This behaviour can be exploited as a way to modulate the morphology and size of the resulted MOF crystals.

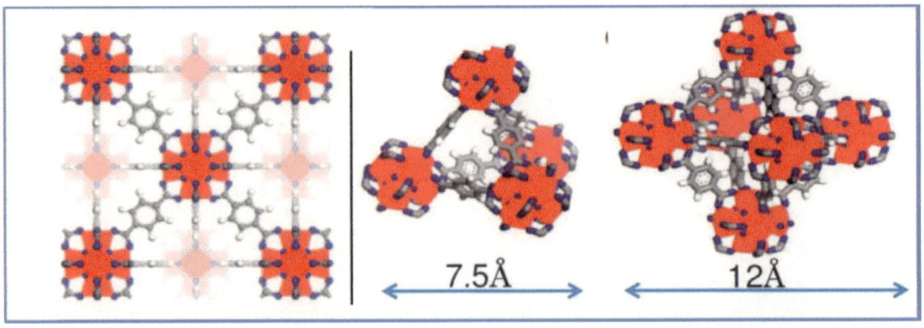

Figure 7.
Schematic illustration of the UiO-66 structure [45].

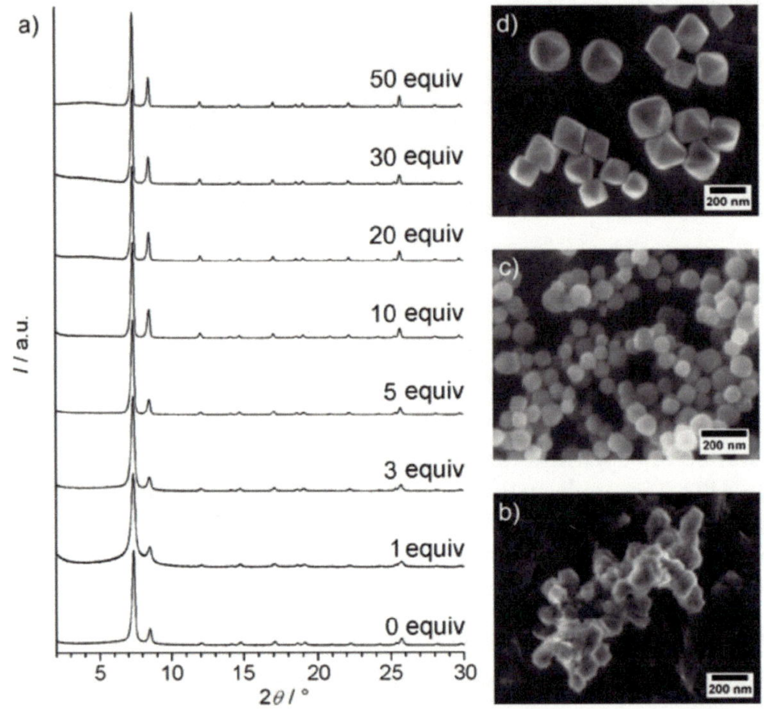

Figure 8.
(a) XRD patterns and (b–d) SEM images of the UiO-66 crystals synthesized with different amounts of benzoic acid additive [46].

4.2 Modulated synthesis of UiO-66 crystals using HF

We have been working on precisely controlled synthesis of the UiO-66 and other MOFs with the aim to optimize their properties and extend their application potentials. We were the first to report a hydrofluoric acid (HF) modulated synthesis of the UiO-66 crystals from reactants $ZrCl_4$ and H_2BDC [47]. In the study, the amount of HF was varied from 1 eq. to 3 eq., where "eq." refers to the molar ratio of HF to $ZrCl_4$ in a synthesis batch. A control sample was also synthesized without the addition of HF but under otherwise identical conditions. **Figure 9** shows the SEM images of the UiO-66 crystals synthesized. As seen, the addition of the HF facilitates growth of the UiO-66 crystals with the mean size of the crystals increasing from ~150 nm to 7 μm with increased HF addition. By controlling the concentration of the reactants, the morphology of crystals changed from truncated cube to cuboctahedron.

To understand the impact of the HF additive, elemental mapping and ^{19}F NMR analysis of the resulted UiO-66 crystals were conducted.

Figure 10 shows the EDS mapping of the synthesized UiO-66 crystals. As seen, without the addition of HF, a small amount of Cl was detected in the crystals even after heat treatment up to 300°C. This somewhat unexpected observation of the Cl is believed to be resulted from structure defects in the synthesized UiO-66 crystals. It is known that the UiO-66 crystal structure contains defects in the form of missing linkers, which would result in an unsaturated framework. To compensate for this charge imbalance of the framework, negatively charged Cl ions (from the $ZrCl_4$ reactant) are "needed" to coordinate with the defective sites.

For the crystals synthesized with the addition of HF, EDS analysis of the crystals revealed noticeable F in the crystals, with no Cl observed. This indicates that, in the presence of stronger electronegative F, the F ions now coordinate to the linker defective sites instead of chlorine ions in the SBUs of the UiO-66 crystals.

This mechanism was further supported by the ^{19}F NMR spectra shown in **Figure 11**. As seen, the spectrum of 3F-UiO-66 heat-treated at 150°C (curve a) contains a strong signal of two partially overlapped peaks cantered at −155 ppm and −156 ppm. These peaks are assigned to the F bonded directly to Zr in the SBUs and the F physisorbed, respectively. When the 3F-UiO-66 is heat-treated at 300°C, as seen in curve (b), the peak at −155 ppm becomes stronger and narrower, whereas the peak at −156 ppm disappears. It indicates that F remains bonded to the Zr in the 3F-UiO-66 framework even after being heat-treated up to 300°C.

The thermostability of the UiO-66 crystals was enhanced after the introduction of fluorine in the framework structure. **Figure 12** shows the TGA curves of the UiO-66 crystals synthesized with different amounts of HF additive. The significant weight loss event observed at ~500°C represents the complete decomposition of the

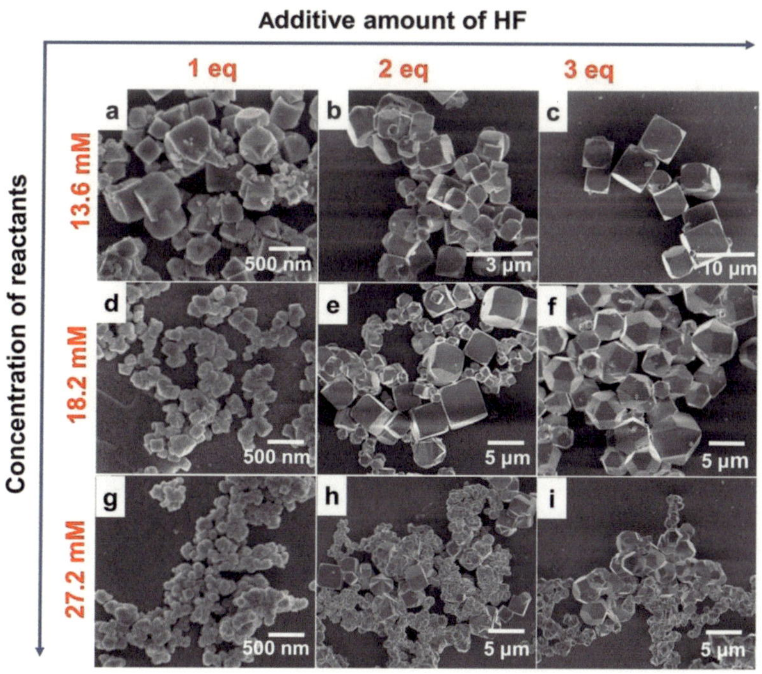

Figure 9.
SEM images of the UiO-66 crystals synthesized with different amounts of HF additive and changing concentrations of the reactants. At $C_{Zr=BDC}$ = 13.6 mM, (a) HF = 1 eq., (b) HF = 2 eq., and (c) HF = 3 eq. At $C_{Zr=BDC}$ = 18.2 mM, (d) HF = 1 eq., (e) HF = 2 eq., and (f) HF = 3 eq. At $C_{Zr=BDC}$ = 27.2 mM, (g) HF = 1 eq., (h) HF = 2 eq., and (i) HF = 3 eq. [47].

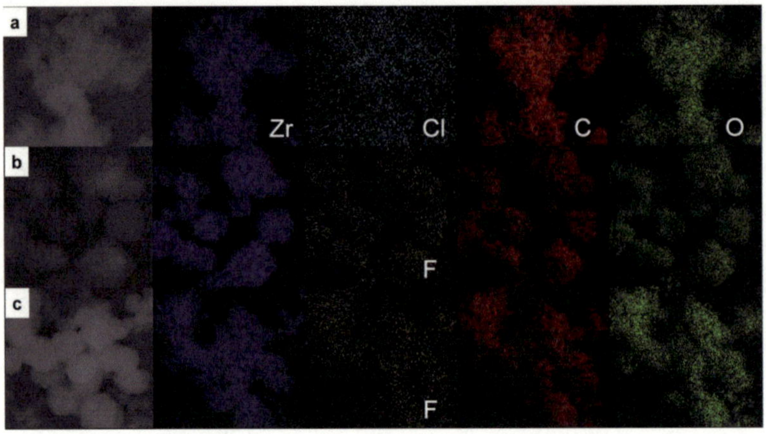

Figure 10.
EDS maps of the UiO-66 crystals synthesized (a) without HF additive, (b) with 3 eq. HF additive, and (c) with 3 eq. HF additive and then heat-treated at 150°C [47].

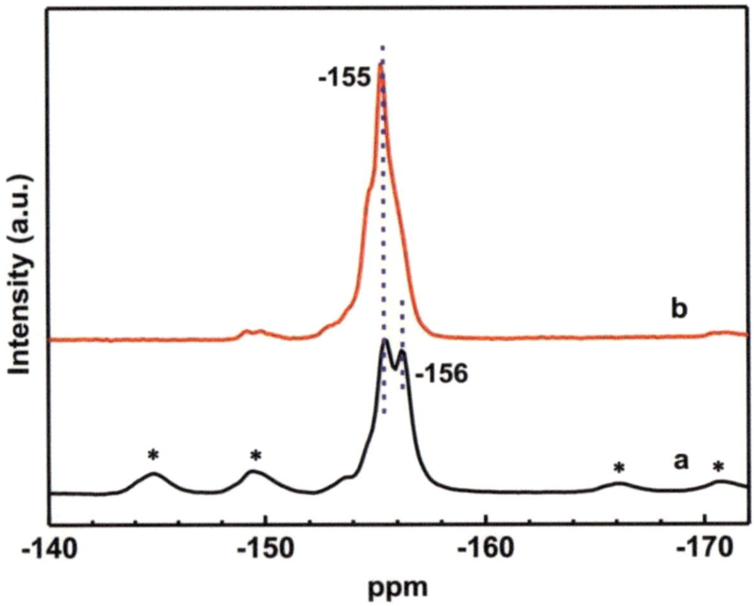

Figure 11.
^{19}F NMR spectra of the 3F-UiO-66 crystals heat-treated at (a) 150°C and (b) 300°C. The asterisks indicate the spinning sidebands [47].

UiO-66 framework. It can be seen that the F introduction has stabilized the framework, as indicated by the progressively increased decomposition temperature. This is attributed to the higher coordination strength of Zr—F in the fluorine-involved crystals than the Zr—Cl or Zr—O in the fluorine-free crystals, resulting in the stabilization of framework structure of the fluorine-involved UiO-66.

Furthermore, as the UiO-66 crystals were modulated from small cube-like morphology to micron-sized cuboctahedron morphology with the addition of the HF during synthesis, their porosity increases [47]. **Figure 13** displays the Ar sorption–desorption isotherms and pore size distributions of the parent and fluorine-involved crystals. As seen, the 3 fluorine-involved crystals (samples c, f, and i in **Figure 9**) exhibit significantly higher Ar adsorption values than the parent crystals

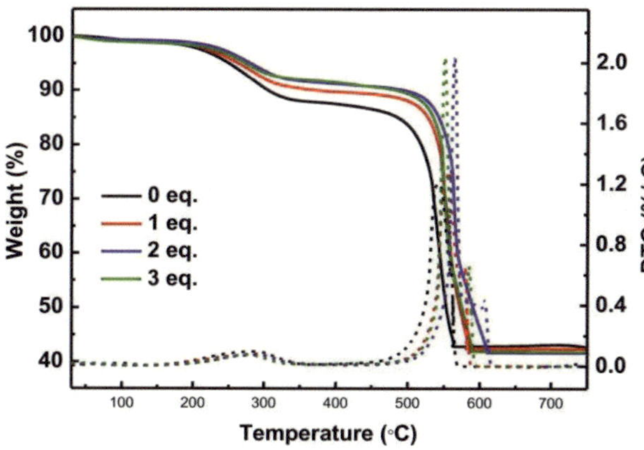

Figure 12.
TGA curves of the UiO-66 crystals synthesized with different amounts of HF additive (C = 18.2 mM) [47].

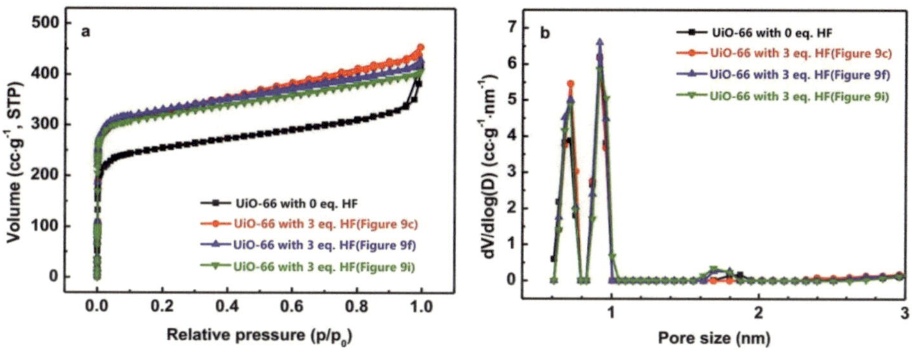

Figure 13.
(a) Ar sorption-desorption isotherms and (b) pore size distributions of the synthesized UiO-66 with various crystal sizes and morphologies [47].

synthesized without HF, and the pore size distributions show the enhanced porosity of the fluorine-involved crystals.

4.3 Modulated synthesis of UiO-66 crystals using solid Cu_2O

In more recent years, nanosized MOFs (also known as nMOFs or NMOFs) have attracted great attentions. nMOFs have short diffusion lengths and high specific surface areas, both are of critical importance in catalysis and sorption, especially in liquid-phase applications [34, 48]. In addition, nMOFs are highly desirable for porous membranes [49, 50], thin film devices [51], and medical applications [52].

We have reported a novel method for the modulated synthesis of nUiO-66 crystals, in which a new type of additive, solid Cu_2O, was used to mediate the synthesis [53]. **Figure 14** shows the TEM images of the UiO-66 crystals synthesized with different amounts of Cu_2O additive. As illustrated, by increasing the amount of Cu_2O additive, the mean sizes of the UiO-66 crystals reduce progressively from ~265 to 40 nm. The crystals maintain an octahedral morphology until when the crystal size is too small to display distinguishable facets and thus appear as spherical particles. The synthesized nano UiO-66 crystals were found to be chemically pure without the contamination of Cu. They have high BET surface area of more than 1100 m^2 g^{-1} and contain rich porosity. **Figure 15** shows the Ar sorption isotherms and pore size distributions of the UiO-66 crystals. Although the amount of micropores

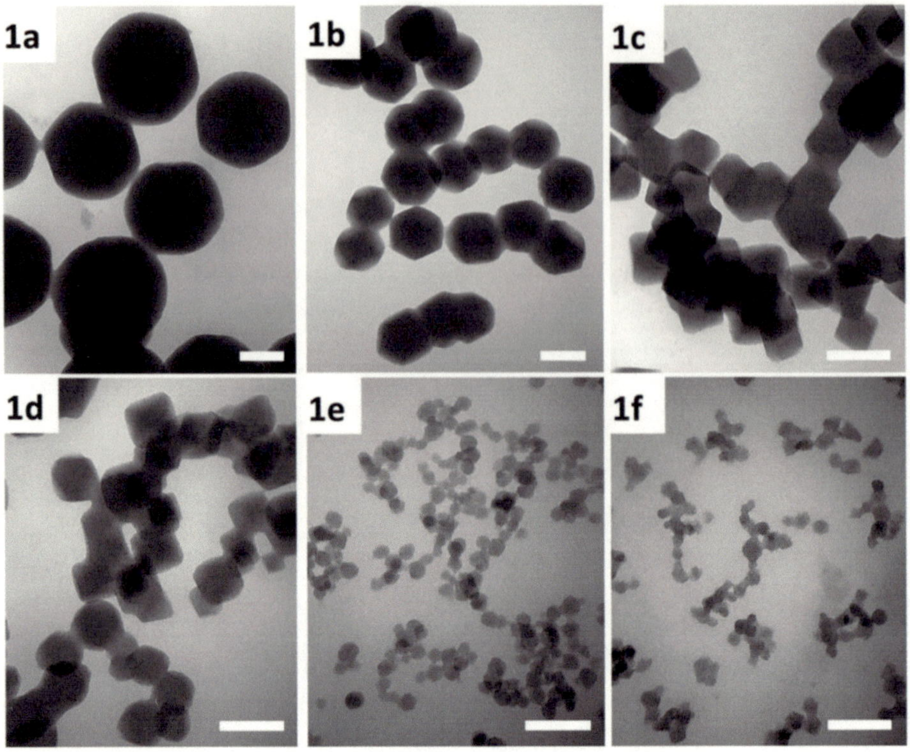

Figure 14.
TEM images of the UiO-66 crystals synthesized with different amounts of Cu_2O additive. Here the molar ratio R = n[Cu]/n[$ZrCl_4$]: (1a) R = 0, (1b) R = 0.2, (1c) R = 0.5, (1d) R = 1.0, (1e) R = 1.2, and (1f) R = 1.5. The scale bar is 100 nm [53].

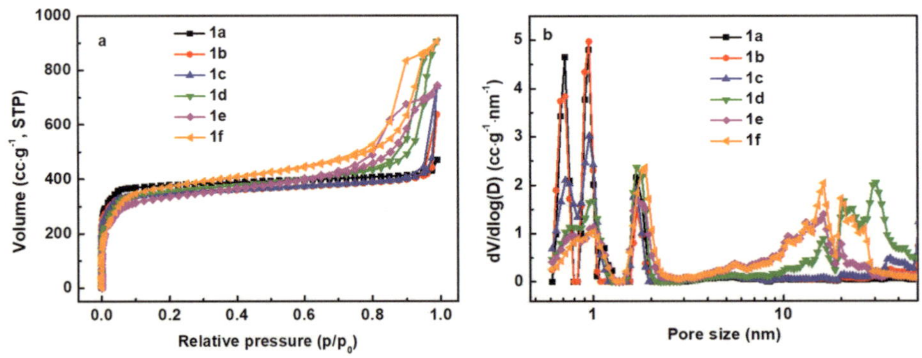

Figure 15.
(a) Ar sorption isotherms and (b) pore size distributions (measured between 0 and 50 nm) of the UiO-66 crystals synthesized with different amounts of Cu_2O additive. Here the molar ratio R = n[Cu]/n[$ZrCl_4$]: (1a) R = 0, (1b) R = 0.2, (1c) R = 0.5, (1d) R = 1.0, (1e) R = 1.2, and (1f) R = 1.5 [53].

between 6 Å and 10 Å decreased gradually with the decreasing crystal size, noticeable quantity of mesopores of 10–30 nm became prevailed, demonstrating the dual micro- and mesoporosity of the nano UiO-66 crystals.

During the Cu_2O-modulated synthesis, the added solid Cu_2O was observed to dissolve gradually in the mother liquor, resulting in the formation of a pale orange-colored solution. **Figure 16** shows the far infrared (FIR) spectra of solid Cu_2O and a number of mixed solutions relevant to the synthesis. As illustrated in the inset

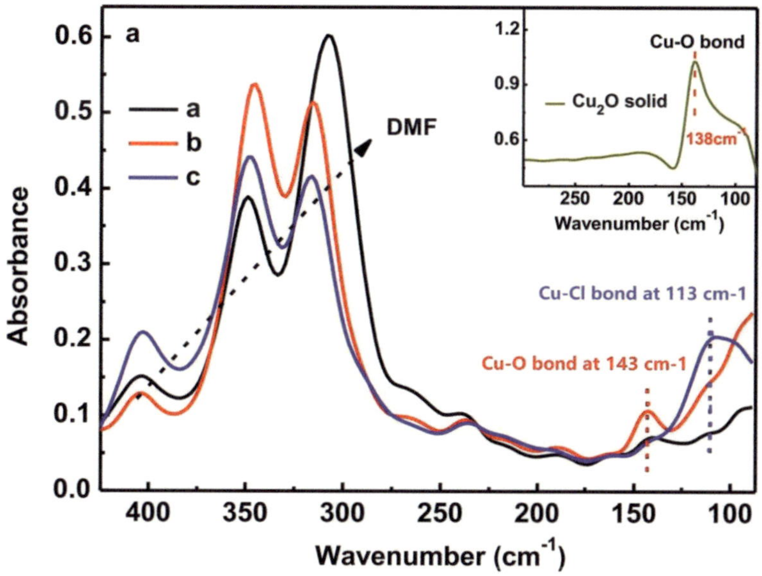

Figure 16.
Far-IR spectra of the solid Cu_2O and a number of mixtures relevant to the UiO-66 synthesis, where the molar ratio of Cu/$ZrCl_4$ is 1.2. (a) $ZrCl_4$/DMF solution, (b) Cu_2O/DMF suspension, and (c) $ZrCl_4$ + Cu_2O/DMF solution. The inset is for solid Cu_2O [53].

of the figure, solid Cu_2O shows the Cu—O absorption band at 138 cm^{-1}. Upon the addition of solid Cu_2O in DMF, the Cu—O bond shifts to 143 cm^{-1} as seen in spectrum b. When the $ZrCl_4$ was added to the suspension of the Cu_2O and DMF, the FIR spectrum (c) shows dramatically decreased Cu—O adsorption band accompanied by the arrival of an intense adsorption band at 113 cm^{-1}. This new band can be ascribed to a shifted stretching vibration of Cu—Cl bonds, which are usually at 109 cm^{-1} for the anionic $[CuCl_2]^-$ complex [54]. The FIR analysis seems to suggest that the dissolution of Cu_2O in the precursor solution of the UiO-66 is driven by the formation of the [Cu—Cl] complexes.

It has also been found [53] that the amount of $ZrCl_4$ in the Cu_2O-modulated synthesis plays a crucial role in the growth of the UiO-66 crystals. The higher the concentration of the $ZrCl_4$, the weaker the Cu_2O's impact on the crystal growth and thus benefits the formation of large size UiO-66 crystals. In contrast, decreasing the concentration of the $ZrCl_4$ favors the production of small UiO-66 crystals, which also accompanied with the loss of crystallinity. It is thus believed that the [Cu—Cl] complex is capable of coordinating with the Zr^{4+}, leading to a decreased coordination of the Zr_6 clusters with BDC linkers, preventing crystal growth.

Additionally, under the modulation of Cu_2O additive, the synthesized UiO-66 crystals not only have reduced sizes, but also enriched structure defects [53]. Both the XRD and TGA analyses of the nano UiO-66 crystals indicate the existence of linker defects and metal cluster defects. **Table 1** presents quantified amounts of the linker missing defects in the synthesized nano UiO-66 crystals. It is seen that the number of linker defects increases with the decreasing crystal size, revealing a more open framework of the nano UiO-66.

In recent years, there have been many reports on the structural defects of the UiO-66 crystals [55–60]. These structural defects are found to relate strongly to the structure stability, adsorption, and catalysts performance of the UiO-66 material.

Sample	Average crystal size of the synthesize nano UiO-66/nm	R	Cu_2O addition	
			x	Formula
1a	265	0	0.42	$Zr_6O_{6.42}(BDC)_{5.58}$
1b	155	0.2	0.50	$Zr_6O_{6.50}(BDC)_{5.50}$
1c	96	0.5	0.84	$Zr_6O_{6.84}(BDC)_{5.16}$
1d	75	1.0	0.84	$Zr_6O_{6.84}(BDC)_{5.16}$
1e	48	1.2	0.84	$Zr_6O_{6.84}(BDC)_{5.16}$
1f	40	1.5	1.05	$Zr_6O_{7.05}(BDC)_{4.95}$

Table 1.
Amounts of linker missing defects in the UiO-66 crystals synthesized with the addition of Cu_2O [53].

Here R represents the molar ratio of $Cu/ZrCl_4$, and x is the number of linker missing defects. The formula represents the defective molecular formula of the UiO-66 crystals, and $Zr_6O_6(BDC)_6$ is the formula of a defect-free dehydroxylated UiO-66.

4.4 In situ synthesis of Ti-doped hybrid UiO-66 crystals

Introducing foreign metal centres into the framework structure to obtain hybrid (i.e. mixed metal) MOFs is a strategy for developing novel MOFs. We have explored the possibility of facile synthesis of Ti^{4+}-doped hybrid UiO-66-*n*Ti MOFs via an in situ crystallization process [61]. In this work, a series of Ti^{4+}-doped UiO-66-*n*Ti (*n* represents the mass fraction of Ti^{4+}) crystals were synthesized following the synthesis procedure of the UiO-66, but with the addition of varying amounts of Ti^{4+} in the

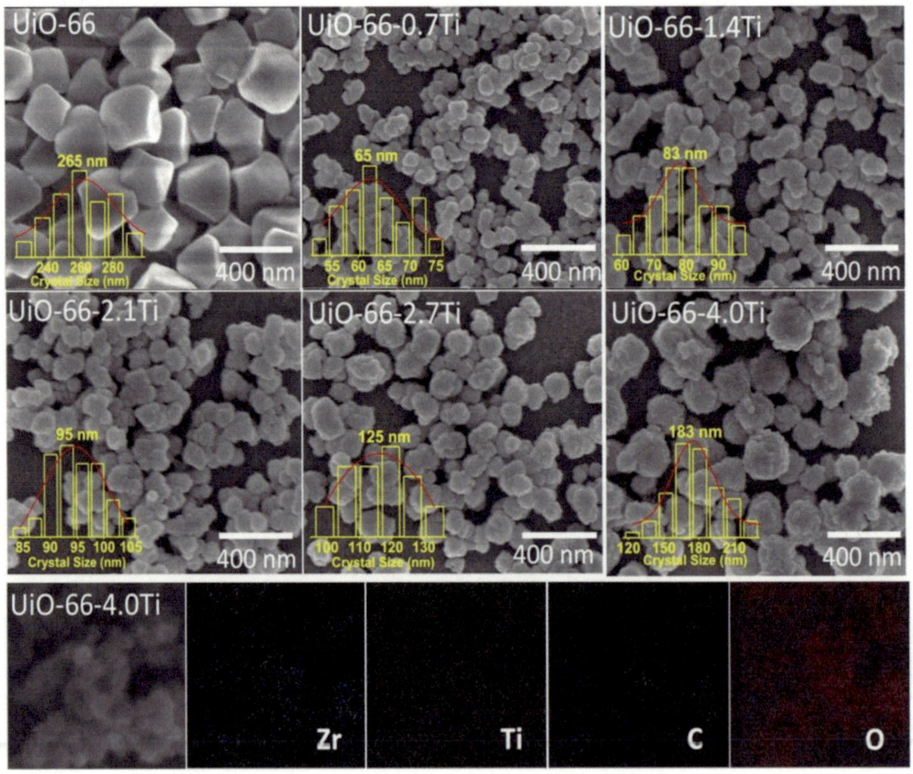

Figure 17.
*SEM images of the UiO-66-*n*Ti MOFs and EDS elemental mappings of UiO-66-4Ti [61].*

form of titanium isopropoxide in the precursor solution. The ICP analysis showed that the concentration of Ti^{4+} increases gradually from around 0.7 to 4.0 wt% in the resulted crystals with increased addition of Ti^{4+} in the synthesis.

Figure 17 shows the SEM images and EDS elemental mapping of the synthesized UiO-66-*n*Ti crystals. It is evident that there is a significant difference on the crystal morphology and size of the UiO-66 crystals as the result of Ti^{4+} doping. The parent UiO-66 contains octahedral crystals of a mean size ~265 nm. With increased Ti^{4+} doping concentration, the UiO-66-*n*Ti crystals change from octahedrons to sphere-like crystals with rougher facets, and their mean crystal sizes decrease sharply from 265 to ~65 nm. This seems to suggest that the doping of Ti^{4+} restraints the growth of the UiO-66-*n*Ti crystals. It is noteworthy that when the Ti^{4+} concentration was further increased, the UiO-66-*n*Ti crystals maintain the sphere-like morphology, but the mean size increases from 65 to 183 nm, unexpectedly.

Figure 18 shows the FT-IR spectra of the parent UiO-66 and doped UiO-66-*n*Ti crystals. As seen, for the parent UiO-66, the Zr—$O_{\mu3\text{-}O}$ stretch in the [Zr—O] clusters is observed at ~682 cm^{-1}, moving towards 664 cm^{-1} with the increase in the Ti^{4+} doping amounts, indicating the structural change in the Ti^{4+}-doped UiO-66-*n*Ti crystals. As reported in the literatures [32, 62], the MIL-125(Ti) and NH_2-MIL-125(Ti) MOFs are built from coordination between Ti_8 clusters and H_2BDC-type organic linkers. Both structures display a FT-IR band at between 600 and 700 cm^{-1} which resulted from (O—Ti—O) vibrations, and for the NH_2-MIL-125(Ti), the (O—Ti—O) vibrations occur at ~642 cm^{-1}. We thus believe that the red shift of the Zr—$O_{\mu3\text{-}O}$ stretch in the UiO-66-*n*Ti crystals could be caused by the incorporation of the Ti^{4+} in the [Zr—O] clusters.

Figure 19 shows the EXAFS spectra of the parent UiO-66 and doped UiO-66-2.7Ti crystals. It further demonstrates the alternation in coordination state around the Zr^{4+} caused by the Ti^{4+} doping. As seen, the peak at ~1.7 Å that relates to the Zr—O bonds shifts to 1.6 Å after Ti^{4+} doping. Additionally, the peak at around 3.0 Å

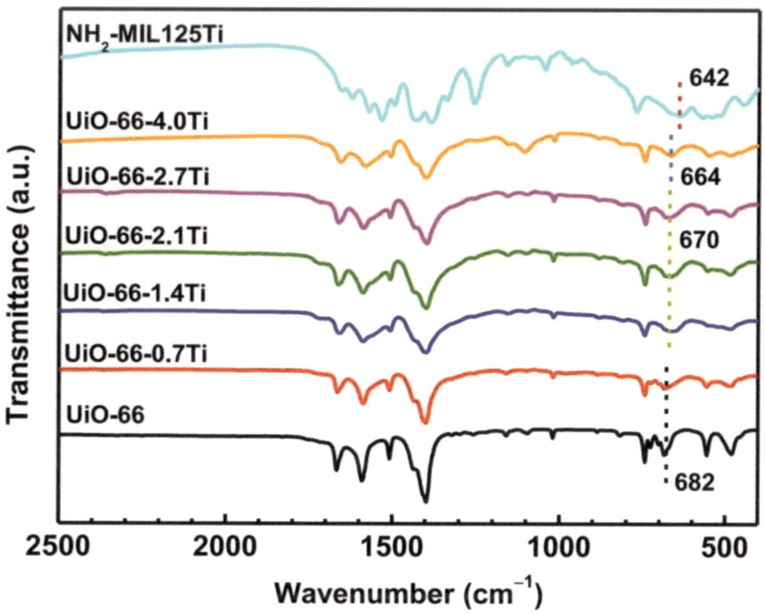

Figure 18.
*FT-IR spectra of the parent UiO-66, doped UiO-66-*nTi *(n = 0.7, 1.4, 2.1, 2.7, and 4.0), and NH_2-MIL-125(Ti) crystals [61].*

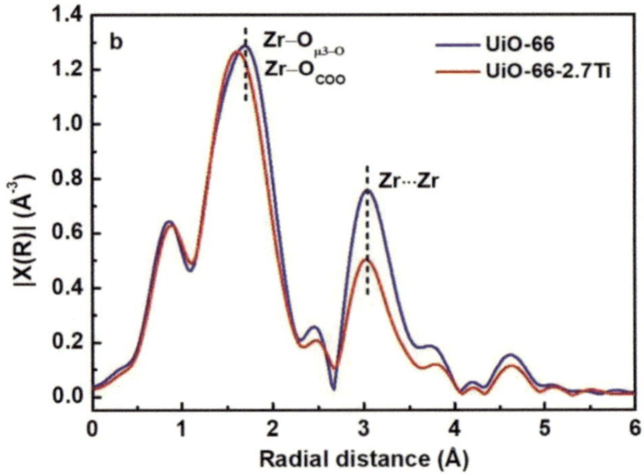

Figure 19.
EXAFS spectra of the parent UiO-66 and the UiO-66-2.7Ti crystals [61].

that represents the Zr···Zr bonds connecting through the $O_{\mu 3\text{-}O}$ or $O_{\mu 3\text{-}OH}$ decreases noticeably after doping, likely due to the formation of Zr···Ti coordination.

The synthesized UiO-66-*n*Ti crystals have been found to be highly selective and efficient in removing anionic dye, the Congo red, from water. The doped UiO-66-2.7Ti (i.e. with 2.72 wt% Ti^{4+} incorporation) exhibited the highest adsorption capacity of 979 mg g^{-1}, 3.6 times of the parent UiO-66 crystals.

5. Conclusions

MOFs, with their inherent high surface areas, uniform, versatile, and tuneable pores, and tailorable physicochemical properties, have been projected to have a broad range of potential applications in various areas. Since the discovery of the first MOF structure in the 1990s, hundreds of new MOFs have been synthesized successfully.

In the past two decades, much attention has been paid to develop methods to control the morphology and size of MOF crystals, in order to tailor the structure properties and performance of MOF materials. Such research efforts have resulted in the development of effective synthesis methods which are capable of modulation of morphology and size of MOF crystals through different mechanisms, including the deprotonation regulation, coordination modulation, and surfactant modulation synthesis. This review chapter has highlighted a number of successful synthesis routes using, for example, the carboxylic acids, HF, and solid Cu_2O additives to modulate the MOF crystal formation and growth. It has also demonstrated the ability of doping to create hybrid MOFs with improved properties and functionalities.

It remains a significant challenge on how to best fully utilize the unique set of properties and functionalities of the chemically versatile MOFs and realize their application potentials in various identified areas. While there have been significant achievements on the structure optimization and performance development of MOFs to date, it is anticipated that further progress can be made with more fundamental research and industrial application of MOFs.

Acknowledgements

This work was supported by the National Natural Science Foundation of China (Grant no. 21236008).

Author details

Yitong Han[1,2], Hong Yang[3] and Xinwen Guo[2]*

1 Dalian Research Institute of Petroleum and Petrochemicals, SINOPEC, Dalian, China

2 State Key Laboratory of Fine Chemicals, PSU-DUT Joint Centre for Energy Research, School of Chemical Engineering, Dalian University of Technology, Dalian, China

3 Department of Mechanical Engineering (M050), The University of Western Australia, Perth, WA, Australia

*Address all correspondence to: guoxw@dlut.edu.cn

References

[1] Sing KS. Reporting physisorption data for gas/solid systems with special reference to the determination of surface area and porosity (recommendations 1984). Pure and Applied Chemistry. 1985;**57**:603-619

[2] Cundy CS, Cox PA. The hydrothermal synthesis of zeolites: History and development from the earliest days to the present time. Chemical Reviews. 2003;**103**:663-701

[3] Xia YD, Yang ZX, Zhu YQ. Porous carbon-based materials for hydrogen storage: Advancement and challenges. Journal of Materials Chemistry A. 2013;**1**:9365-9381

[4] Wales DJ, Grand J, Ting VP, Burke RD, Edler KJ, Bowen CR, et al. Gas sensing using porous materials for automotive applications. Chemical Society Review. 2015;**44**:4290-4321

[5] Sun MH, Huang SZ, Chen LH, Li Y, Yang XY, Yuan ZY, et al. Applications of hierarchically structured porous materials from energy storage and conversion, catalysis, photocatalysis, adsorption, separation, and sensing to biomedicine. Chemical Society Review. 2016;**45**:3479-3563

[6] Chen YQ, Chen KW, Bai H, Li L. Electrochemically reduced graphene porous material as light absorber for light-driven thermoelectric generator. Journal of Materials Chemistry. 2012;**22**:17800-17804

[7] Wang BJ, Prinsen P, Wang HZ, Bai ZS, Wang HL, Luque R, et al. Macroporous materials: Microfluidic fabrication, functionalization and applications. Chemical Society Review. 2017;**46**:855-914

[8] Kitagawa S, Kitaura R, Noro S. Functional porous coordination polymers. Angewandte Chemie, International Edition. 2004;**43**:2334-2375

[9] Ockwig NW, Delgado-Friedrichs O, O'keeffe M, Yaghi OM. Reticular chemistry: Occurrence and taxonomy of nets and grammar for the design of frameworks. Accounts of Chemical Research. 2005;**38**:176-182

[10] Zhu QL, Xu Q. Metal-organic framework composites. Chemical Society Review. 2014;**43**:5468-5512

[11] Jia JH, Lin X, Wilson C, Blake AJ, Champness NR, Hubberstey P, et al. Twelve-connected porous metal-organic frameworks with high H_2 adsorption. Chemical Communications. 2007;**8**:840-842

[12] Rosi NL, Eckert J, Eddaoudi M, Vodak DT, Kim J, O'Keeffe M, et al. Hydrogen storage in microporous metal-organic frameworks. Science. 2003;**300**:1127-1129

[13] Férey G, Mellot-Draznieks C, Serre C, Millange F, Dutour J, Surblé S, et al. A chromium terephthalate-based solid with unusually large pore volumes and surface area. Science. 2005;**309**:2040-2042

[14] Furukawa H, Cordova KE, O'Keeffe M, Yaghi OM. The chemistry and applications of metal-organic frameworks. Science. 2013;**341**:1230444

[15] Queen WL, Brown CM, Britt DK, Zajdel P, Hudson MR, Yaghi OM. Site-specific CO_2 adsorption and zero thermal expansion in an anisotropic pore network. Journal of Physical Chemistry C. 2011;**115**:24915-24919

[16] Xue DX, Belmabkhout Y, Shekhah O, Jiang H, Adil K, Cairns AJ, et al. Tunable rare earth Fcu-MOF platform: Access to adsorption kinetics driven gas/vapor separations via pore

size contraction. Journal of American Chemical Society. 2015;**137**:5034-5040

[17] Belmabkhout Y, Guillerm V, Eddaoudi M. Low concentration CO_2 capture using physical adsorbents: Are metal–organic frameworks becoming the new benchmark materials? Chemical Engineering Journal. 2016;**296**:386-397

[18] Øien-Ødegaard S, Bouchevreau B, Hylland K, Wu LP, Blom R, Grande C, et al. UiO-67-type metal-organic frameworks with enhanced water stability and methane adsorption capacity. Inorganic Chemistry. 2016;**55**:1986-1991

[19] Guo YX, Feng X, Han TY, Wang S, Lin ZG, Dong YP, et al. Tuning the luminescence of metal-organic frameworks for detection of energetic heterocyclic compounds. Journal of American Chemical Society. 2014;**136**:15485-15488

[20] Nickerl G, Senkovska I, Kaskel S. Tetrazine functionalized zirconium MOF as an optical sensor for oxidizing gases. Chemical Communications. 2015;**51**:2280-2282

[21] Yang XQ, Yan DP. Long-afterglow metal-organic frameworks: Reversible guest-induced phosphorescence tunability. Chemical Science. 2016;**7**:4519-4526

[22] Pentyala V, Davydovskaya P, Ade M, Pohle R, Urban G. Metal-organic frameworks for alcohol gas sensor. Sensors and Actuators B. 2016;**222**:904-909

[23] Wu XL, Yang C, Ge J, Liu Z. Polydopamine tethered enzyme/metal-organic framework composites with high stability and reusability. Nanoscale. 2015;**7**:18883-18886

[24] Choi KM, Na K, Somorjai GA, Yaghi OM. Chemical environment control and enhanced catalytic performance of platinum nanoparticles embedded in nanocrystalline metal-organic frameworks. Journal of American Chemical Society. 2015;**137**:7810-7816

[25] Carson F, Martínez-Castro E, Marcos R, Miera GG, Jansson K, Zou XD, et al. Effect of the functionalisation route on a Zr-MOF with an Ir-NHC complex for catalysis. Chemical Communications. 2015;**51**:10864-10867

[26] Yamada T, Otsubo K, Makiura R, Kitagawa H. Designer coordination polymers: Dimensional crossover architectures and proton conduction. Chemical Society Reviews. 2013;**42**:6655-6669

[27] Liu JX, Zhou WC, Liu JX, Howard I, Kilibarda G, Schlabach S, et al. Photoinduced charge-carrier generation in epitaxial MOF thin films: High efficiency as a result of an indirect electronic band gap? Angewandte Chemie International Edition. 2015;**54**:7441-7445

[28] Bao T, Zhang J, Zhang WP, Chen ZL. Growth of metal-organic framework HKUST-1 in capillary using liquid-phase epitaxy for open-tubular capillary electrochromatography and capillary liquid chromatography. Journal of Chromatography A. 2015;**1381**:239-246

[29] Qiu S, Xue M, Zhu GS. Metal-organic framework membranes: From synthesis to separation application. Chemical Society Reviews. 2014;**43**:6116-6140

[30] Rubio-Martinez M, Avci-Camur C, Thornton AW, Imaz I, Maspoch D, Hill MR. New synthetic routes towards MOF production at scale. Chemical Society Review. 2017;**46**:3453-3480

[31] Tao AR, Habas S, Yang PD. Shape control of colloidal metal nanocrystals. Small. 2008;**4**:310-325

[32] Hu S, Liu M, Li KY, Zou Y, Zhang AF, Song CS, et al. Solvothermal synthesis of NH_2-MIL-125(Ti) from circular plate to octahedron. CrystEngComm. 2014;**16**:9645-9650

[33] Zahn G, Zerner P, Lippke J, Kempf FL, Lilienthal S, Schröder CA, et al. Insight into the mechanism of modulated syntheses: In situ synchrotron diffraction studies on the formation of Zr-Fumarate MOF. CrystEngComm. 2014;**16**:9198-9207

[34] Tsuruoka T, Furukawa S, Takashima Y, Yoshida K, Isoda S, Kitagawa S. Nanoporous nanorods fabricated by coordination modulation and oriented attachment growth. Angewandte Chemie, International Edition. 2009;**48**:4739-4743

[35] Diring S, Furukawa S, Takashima Y, Tsuruoka T, Kitagawa S. Controlled multiscale synthesis of porous coordination polymer in nano/micro regimes. Chemistry of Materials. 2010;**22**:4531-4538

[36] Umemura A, Diring S, Furukawa S, Uehara H, Tsuruoka T, Kitagawa S. Morphology design of porous coordination polymer crystals by coordination modulation. Journal of the American Chemical Society. 2011;**133**:15506-15513

[37] Jiang DM, Mallat T, Krumeich F, Baiker A. Polymer-assisted synthesis of nanocrystalline copper-based metal-organic framework for amine oxidation. Catalysis Communications. 2011;**12**:602-605

[38] Cai XC, Lin J, Pang ML. Facile synthesis of highly uniform Fe-MIL-88B particles. Crystal Growth & Design. 2016;**16**:3565-3568

[39] Yang JM, Qi ZP, Kang YS. Effect of additives on morphology and size and gas adsorption of SUMOF-3 microcrystals. Microporous and Mesoporous Materials. 2016;**222**:27-32

[40] Li WH, Zhang AF, Jiang X, Chen C, Liu ZM, Song CS, et al. Low temperature CO_2 methanation: ZIF-67-derived co-based porous carbon catalysts with controlled crystal morphology and size. ACS Sustainable Chemical Engineer. 2017;**5**:7824-7831

[41] Ma MY, Zacher D, Zhang XN, Fisher RA, Metzler-Nolte NA. Method for the preparation of highly porous, nanosized crystals of isoreticular metal-organic frameworks. Crystal Growth & Design. 2011;**11**:185-189

[42] Yang JM, Liu Q, Kang YS. Controlled growth and gas sorption properties of IRMOF-3 nano/microcrystals. Dalton Transactions. 2014;**43**:16707-16712

[43] Liu Q, Jin LN, Facile Fabrication SWY. Adsorption property of a nano/microporous coordination polymer with controllable size and morphology. Chemical Communication. 2012;**48**:8814-8816

[44] Pan YC, Heryadi D, Zhou F, Zhao L, Lestari G, Su HB, et al. Tuning the crystal morphology and size of zeolitic imidazolate framework-8 in aqueous solution by surfactants. CrystEngComm. 2011;**13**:6937-6940

[45] Cavka JH, Jakobsen S, Olsbye U, Guillou N, Lamberti C, Bordiga S, et al. A new zirconium inorganic building brick forming metal organic frameworks with exceptional stability. Journal of the American Chemical Society. 2008;**130**:13850-13851

[46] Schaate A, Roy P, Godt A, Lippke J, Waltz F, Wiebcke M, et al. Modulated synthesis of Zr-based metal-organic frameworks: From nano to single

crystals. Chemistry— A European Journal. 2011;**17**:6643-6651

[47] Han YT, Liu M, Li KY, Zuo Y, Wei YX, Xu ST, et al. Facile synthesis of morphology and size-controlled zirconium metal-organic framework UiO-66: The role of hydrofluoric acid in crystallization. CrystEngComm. 2015;**17**:6434-6440

[48] Díaz-García M, Mayoral Á, Díaz I, Sánchez-Sánchez M. Nanoscaled M-MOF-74 materials prepared at room temperature. Crystal Growth & Design. 2014;**14**:2479-2487

[49] Huang AS, Wang NY, Caro J. Seeding-free synthesis of dense zeolite FAU membranes on 3-aminopropyltriethoxysilane-functionalized alumina supports. Journal of Membrane Science. 2012;**389**:272-279

[50] Mao YY, Cao W, Li JW, Sun LW, Peng XS. HKUST-1 membranes anchored on porous substrate by hetero MIL-110 nanorod array seeds. Chemistry— A European Journal. 2013;**19**:11883-11886

[51] Stassen I, Campagnol N, Fransaer J, Vereecken P, Vos DD, Ameloot R. Solvent-free synthesis of supported ZIF-8 films and patterns through transformation of deposited zinc oxide precursors. CrystEngComm. 2013;**15**:9308-9311

[52] deKrafft KE, Boyle WS, Burk LM, Zhou OZ, Lin WB. Zr- and Hf-based nanoscale metal-organic frameworks as contrast agents for computed tomography. Journal of Materials Chemistry. 2012;**22**:18139-18144

[53] Han YT, Liu M, Li KY, Sun Q, Song CS, Zhang GL, et al. Cu_2O mediated synthesis of metal-organic framework UiO-66 in nanometer scale. Crystal Growth & Design. 2017;**17**:685-692

[54] Nakamoto K. Infrared and Raman Spectra of Inorganic and Coordination Compounds. 3rd ed. New York: Wiley; 1978. pp. 115-137

[55] Fang ZL, Bueken B, Vos DED, Fischer RA. Defect-engineered metal–organic frameworks. Angewandte Chemie, International Edition. 2015;**54**:7234-7254

[56] Vandichel M, Hajek J, Ghysels A, Vos AD, Waroquier M, Speybroeck VV. Water coordination and dehydration processes in defective UiO-66 type metal organic frameworks. CrystEngComm. 2016;**18**:7056-7069

[57] Shearer GC, Chavan S, Bordiga S, Svelle S, Olsbye U, Lillerud KP. Defect engineering: Tuning the porosity and composition of the metal-organic framework UiO-66 via modulated synthesis. Chemistry of Materials. 2016;**28**:3749-3761

[58] Vandichel M, Hajek J, Vermoortele F, Waroquier M, Vos DED, Speybroeck V. Active site engineering in UiO-66 type metal–organic frameworks by intentional creation of defects: A theoretical rationalization. CrystEngComm. 2015;**17**:395-406

[59] Thornton AW, Babarao R, Jain A, Trousselet F, Coudert FX. Defects in metal-organic frameworks: A compromise between adsorption and stability? Dalton Transactions. 2016;**45**:4352-4359

[60] Canivet J, Vandichel M, Farrusseng D. Origin of highly active metal-organic frameworks catalysts: Defects? Defects! Dalton Transactions. 2016;**45**:4090-4099

[61] Han YT, Liu M, Li KY, Sun Q, Zhang WS, Song CS, et al. In situ synthesis of titanium doped hybrid metal-organic framework UiO-66 with enhanced adsorption capacity

for organic dyes. Inorganic Chemistry Frontiers. 2017;**4**:1870-1880

[62] Dan-Hardi M, Serre C, Frot T, Rozes L. A new photoactive crystalline highly porous titanium (IV) dicarboxylate. Journal of the American Chemical Society. 2009;**131**:10857-10859

Chapter 6

Laser Floating Zone: General Overview Focusing on the Oxyorthosilicates Growth

Francisco Rey-García, Carmen Bao-Varela and Florinda M. Costa

Abstract

This chapter reviews the laser floating zone (LFZ) technique, also known as the laser-heated pedestal growth (LHPG), focusing on the recently produced rare-earth-doped oxyorthosilicate fibers. LFZ has been revealed as a suitable prototyping technique since high-quality crystals can be developed in short time with low consumption of precursor materials in a crucible-free processing that ensures to practically avoid by-products. Moreover, additional advantages are the possibility to treat and melt highly refractory materials together with the easy way for tailoring the final microstructural characteristics and this way the macroscopic physical properties. Thus, refractory rare-earth (RE) doped oxyorthosilicates following the formula RE_2SiO_5 have been recently produced by the LFZ technique for tuning laser emission parameters. The oxyorthosilicates have high chemical stability and allow incorporation of many rare-earth ions yielding different applications, such as laser host materials, gamma ray detectors or scintillators, environmental barrier coatings (EBCs) and waveguides, among others. Thus, different kinds of oxyorthosilicates were produced by the LFZ technique, and the detailed effects of the main processing parameters on crystal's characteristics are discussed in this chapter.

Keywords: laser floating zone, crystal growth, oxyorthosilicates, rare earths, single crystal, polycrystalline ceramics

1. Introduction

The production of high-quality silicate-based single crystals is mainly accomplished by solid state and Czochralski methods ([1–9] and references therein). However, these methods require several amounts of material and the use of crucibles that can introduce external contamination. Moreover, expensive crucibles such as platinum or iridium and special atmospheres are usually necessary when the desired materials are refractory or their chemical reactivity can negatively affect the phase development. So, all these restrictions together with long processing time considerably increase the production costs, being not the most suitable approach for materials prototyping.

The micro-pulling down (μPD) and the floating zone (FZ) are alternative techniques to grow crystalline fibers from a melt [10–12]. The μPD technique is suitable for prototyping; however, the melt is continuously in contact with crucible, being also applied low pulling rates. Concerning the FZ technique, the preform materials with cylindrical geometry (rods) are placed inside a mirror-like concave chamber provided with halogen lamps that allow the material melting in a small region. These techniques are complex and limited by the melting point of the materials.

The laser floating zone (LFZ) technique is similar to the FZ; however, a laser beam, guided into a closed chamber through a ZnSe window, is used to melt the top of a feed rod precursor material. Afterwards, a seed rod is immersed into the molten zone and pulled at a controlled pulling rate [13, 14]. This technique presents many advantages when compared with standard growth methods, namely the growth at high pulling rates, the synthesis of materials with very high melting temperature [15–17] and the most important one: it is a crucible-free process, thus avoiding any external related contamination [14, 18]. This way high purity crystals can be obtained in a short time from a small amount of raw material and minor energy consumption. Moreover, being a nonequilibrium process, metastable phases can be developed from the solid/liquid interface due to the very high thermal gradients [19]. **Figure 1** puts in evidence a scheme of the LFZ process.

The LFZ equipment comprises a laser system coupled to a reflective optical setup, composed by a *reflaxicon*, and a plane mirror and a parabolic mirror. The term *reflaxicon* was introduced in 1970 by Edmonds [20], and it describes a two-stage pair of reflective linear axicon surfaces (**Figures 1** and **2**). As Edmonds [20] did not consider nonlinear axicons, all the applications that he proposed were afocal in nature. This reflective device essentially consists of a primary conical mirror and a larger secondary conical mirror coaxially located with respect to the primary. The function of this device is to convert a solid light beam into a hollow one in an essentially lossless manner (except for absorption at the mirror surface and other similar phenomena). This device is similar to the one patented by Martin in 1948 [21]. A circular crown-shaped laser beam is obtained by the mirror aiming to produce a uniform radial heating. In the LFZ process, after the *reflaxicon*, the plane mirror setted up 45° allows the laser beam reflection to the vertical position in the direction of the spherical or parabolic mirror. The rod precursor defines the crown size of the fiber produced, and a floating zone configuration is obtained [14, 18]. It must be noted that mirrors have a hole in their centers allowing feed and seed holders be placed in the optical axis [14]. Furthermore, the use of a closed chamber allows the growth under controlled atmosphere [22]. Additionally, the growth is controlled by a camera video system focused into the floating zone area allowing to observe the molten zone and particularly the melting and the crystallization interface [18].

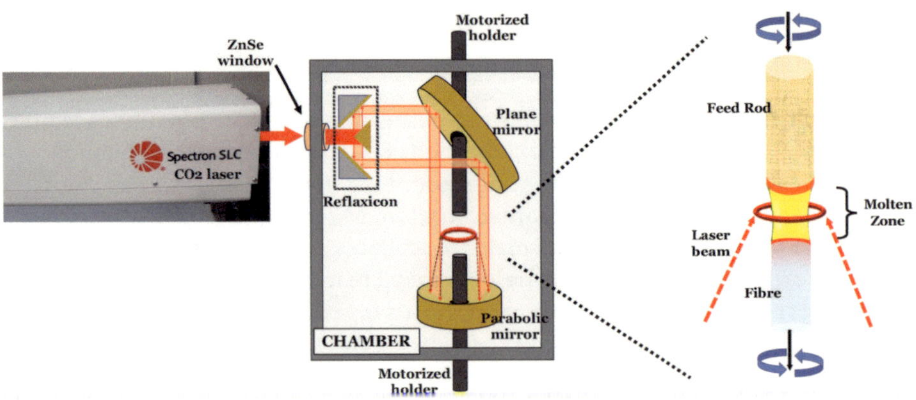

Figure 1.
Laser floating zone setup, highlighting the molten zone.

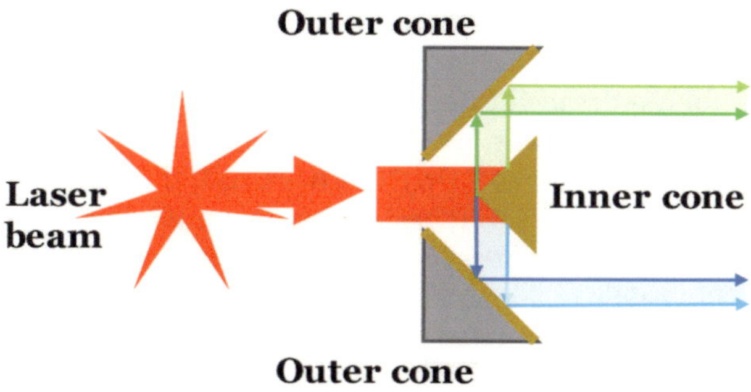

Figure 2.
Reflaxicon *performance noting light guiding with different colors to enhance comprehension.*

Andreeta et al. [14] have reported a good description of the LFZ systems, putting in evidence their modifications of the LFZ systems over the years. A highlight is made to all processes developed to control/modify the temperature gradient at the solid/liquid interface localized between the molten zone and feed/seed rods (**Figure 1**). Noteworthy is the method of electrically assisted laser floating zone (EALFZ) [23, 24] that explores a new mechanism for controlling the solidification process by applying an electric current through the solid/liquid interface. In the presence of an electric current, the solute transport depends not only on the local solute gradient but also on the electromigration of solute that will modify the composition and the characteristic length scale of the solute diffusion field ahead of crystals. The application of an electrical current strongly modified phase development, crystal shapes and effective distribution coefficients. The enhancement of nucleation rate and of the driving force for ion migration along the fiber axis promotes a strong increase in the grain alignment and consequently on the physical properties [25, 26].

The LFZ method enables the development of high-quality light emitting materials and transparent conductive oxides with rare-earth doping [15, 27, 28] along with the growth of complex ceramic materials such as thermoelectric oxide materials [25, 29], high-temperature ceramic superconductors [19, 23, 26] and eutectic oxides [13, 30]. Alongside, by controlling the laser irradiance and the pulling rate, it is possible to determine the crystallization kinetics aiming to obtain highly oriented single or polycrystalline materials, with enhanced physical properties [16, 19, 24].

2. Oxyorthosilicates

Considering the LFZ characteristics, the focus of this work is on oxyorthosilicates, following the formula RE_2SiO_5 (RE = Gd, Lu and Nd). These silicates are highly refractory materials, and despite they present high chemical stability, it is possible to incorporate high concentrations of rare earth ions. Thus, they have attracted the attention of the researchers for long time, yielding different applications [2, 31]. Indeed, since their discovery by Toropov et al. [32], they have been applied as laser host materials [1–3, 9, 33–35], gamma ray detectors or scintillators [36], environmental barrier coatings (EBCs) [37] or, even, as waveguides [38]. The interest in this kind of compounds arose after the first study carried out by Hopkins et al. [39] based on the growth of the rare earth oxyapatites. However, it is important to highlight that these high refractory silicates were grown by the CZ method [1–5, 35–37, 39–41]. For example, Ryba-Romanovski et al. [1] developed solid state yellow lasers based on $(Lu_xGd_{1-x})_2SiO_5$:Sm crystals by CZ, while Wu et al. [41] have recently grown Cu co-doped Ce:Lu_2SiO_5 crystals at 1.5 mm h^{-1} using a iridium crucible under nitrogen atmosphere for application as scintillators. Furthermore, other approaches for oxyorthosilicates production have been employed during the last years using different methods, like pulsed laser deposition [42], sol–gel [43] or solid-state diffusional process [44]. On the other hand, their production by the LFZ technique is more recent [18, 45–47]; besides, there are two experiments performed in the 1980s decade by de la Fuente et al. [33] and Black et al. [34], who produced $GdNdSiO_5$ and $7Gd_2O_3$•$9SiO_2$:Nd materials applying a high laser power (~185 W).

Aiming to develop new laser materials, Rey-García et al. [18, 46, 47] have recently produced a sort of gadolinium-lutetium oxyorthosilicate materials at low laser powers (<100 W) and pulling rates two to three times faster than those used by Czochralski or other standard methods (1-3 mm h^{-1}). Thus, transparent fibers of Gd_2SiO_5 [18], $(Gd_{0.3}Lu_{0.7})_2SiO_5$ [46] and 5 mol% Y^{3+}:$(Gd_{0.3}Lu_{0.7})_2SiO_5$ [47] have been obtained at 10 mm h^{-1} in air under atmospheric pressure presenting high crystallinity degree. These single crystals present excellent photonic properties that make them useful to be employed as laser host materials due to the Gd^{3+} charger transfer band (CTB) that favors the transfer processes with $4f^7$ transitions from the $^8S_{7/2}$ ground state to energy levels of the dopant element [48–50]. Regarding compositional aspects, despite the similarity observed in these single crystals produced by LFZ, the growing processes based on stoichiometric mixtures of Lu_2O_3 or Nd_2O_3 with SiO_2 bring considerable deviations on the phases diagram associated to crystallization paths that can induce materials evaporation or phases rearrangement [45]. Likewise, the LFZ suitability could be sometimes compromised by precursor's properties, nominal compositions or growing conditions. Consequently, remarkable crack formation can be developed due to internal stress mainly induced by the biaxial character of these silicates and the experimental growing parameters [18].

Summarizing, this chapter will highlight the suitability of the LFZ technique on developing compact and miniaturized crystals envisaging new photonic devices, through the production of low volume bulks with an appropriate geometry based on oxyorthosilicates. The optical fundaments of the LFZ technique together with practical aspects relating to oxyorthosilicates production will be described before showing the microstructural and photonic properties of the materials produced. The idea is to demonstrate the LFZ technique as a suitable, time-saving and economic process for laser materials prototyping compared with traditional techniques [14, 40].

3. Experimental

The extrusion process is the most common way to prepare the precursor rod cylinders for the LFZ process, since it is a simple method, not requiring special equipment or additional hands. Thus, the commercial raw oxide powders are mixed, according to the desired stoichiometry, and reduced into grain size with an agate ball mill for 2 hours at 200 rpm. The purity of the precursors depends on the desired application, being used, in this study, powders of 5–6 N of purity since photonics applications are envisaging. Aiming to bind the powder mixture for extrusion process, polyvinyl alcohol (PVA, 0.1 g ml^{-1}) is added in mashing the powders until a compact and plastic paste is achieved. Then, the obtained clay is extruded into cylindrical rods, with diameters depending on the material applications. In the case of the oxyorthosilicates, diameters ranging 1.5–2.0 mm were selected.

After extrusion, the cylindrical rods are dried in air, being ready to be used as feed and seed materials. However, in the LFZ process, single crystals can also be used as the seed rods [14], favoring the formation of a single crystalline fiber. This approach helps laser processing and allows enhancing the structural

Nominal formula	Sample acronym	Power (W)	Obtained composition	Fiber type
Gd_2SiO_5	GSO	72	Gd_2SiO_5	Crystal
$(Lu_{0.1}Gd_{0.9})_2SiO_5$	LGSO-1	67	$(Lu_{0.12}Gd_{0.88})_2SiO_5$	Crystal
$(Lu_{0.3}Gd_{0.7})_2SiO_5$	LGSO-3	58	$(Lu_{0.31}Gd_{0.69})_2SiO_5$	Crystal
$(Lu_{0.5}Gd_{0.5})_2SiO_5$	LGSO-5	64	$(Lu_{0.53}Gd_{0.47})_2SiO_5$	Crystal
Lu_2SiO_5	LSO-10	92	Lu_2SiO_5/Lu2O_3	Eutectics
Nd_2SiO_5	NSO-10	69	Nd_2SiO_5/$Nd_{9.33}(SiO_4)_6O_2$	Biphasic

All fibers have diameters of 1.5 mm for all samples except Lu_2SiO_5 that have 2 mm.

Table 1.
Oxyorthosilicates fibers grown at 10 mm h^{-1} in air by LFZ.

characteristics of the single crystal fiber produced. However, this approach was declined for oxyorthosilicates due to their high melting points.

The LFZ equipment employed for the oxyorthosilicates growth comprise a 200 W CO_2 laser (Spectron, GSI group) coupled to a reflective optical set-up described in the previous section, Optical Fundaments. Once seed and feed fibers are placed on the respective holders, a fast growth process was performed aiming to obtain dense precursor materials. This densification step occurred at 100 mm h^{-1} pulling rate and promotes the PVA decomposition, the formation of the desired phases and enhances the rods mechanical properties. A molten zone is formed by irradiating the densified rods with the CO_2 laser. The fibers were grown in descendent direction from this molten region at 10 mm h^{-1} in air at atmospheric pressure, (crystallization step). Simultaneously during growth, the feed and seed rods rotated at 5 rpm in opposite direction. This procedure favors the mixing of precursors in the melt, homogenizes the temperature of the molten material and contributes to reduce the thermal stresses. In the case of oxyorthosilicate single crystals based on $(Lu_xGd_{1-x})_2SiO_5$ (x = 0–1), the growing process should end by reducing the laser power gradually. This procedure is very important to reduce the thermal stresses and, therefore, avoiding the crack formation.

Table 1 summarizes the experimental conditions to grow oxyorthosilicate fibers at 10 mm h^{-1} in air by LFZ. Considering the Gd_2O_3, Lu_2O_3 and Nd_2O_3 melting points, (2420, 2490 and 2233°C, respectively), the slight variation of the laser power irradiance well matches with this small melting point variation. Despite the experimental conditions are similar, the fibers developed varied from single crystal (GSO and derived silicates) to eutectic (LSO derived) and biphasic (NSO derived) ceramics due to specific characteristic of each phase diagram. Concomitantly, the most remarkable characteristics of the oxyorthosilicates produced in air by LFZ will be described below.

4. Results

Regarding structural properties, the silicates having the formula RE_2SiO_5 are all monoclinic, presenting $P2_1/c$ (Gd_2SiO_5, [18]) or C2/c (Lu_2SiO_5, [45]) space groups depending on the rare earth ions present (**Figure 3**) [46, 51]. This structure provokes their biaxial character that compromises their crystallization, resulting in internal crack formation when LFZ processing is carried out in air and the cooling is not gradually performed [18, 47]. **Figure 4** shows the XRD powder diffractogram of the Gd, Lu and Nd oxyorthosilicates presented in this chapter. It should be noted

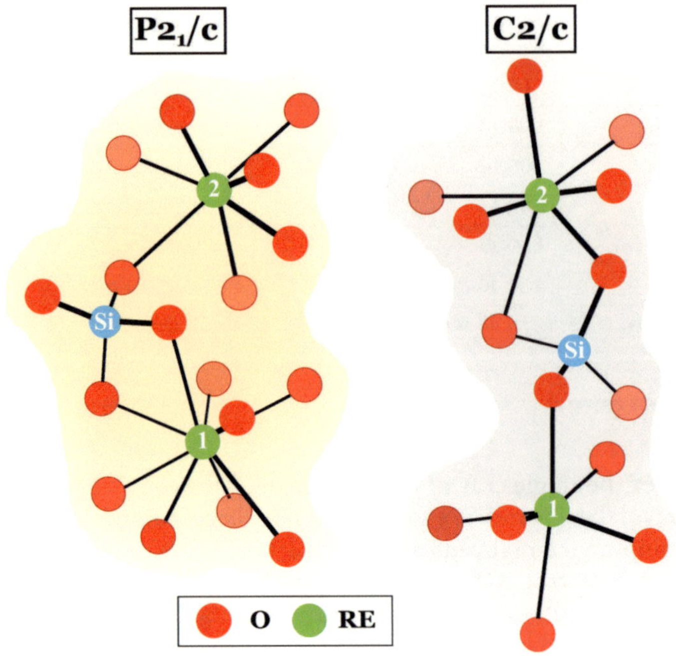

Figure 3.
Structural scheme of $P2_1/c$ and C2/c spatial groups [46].

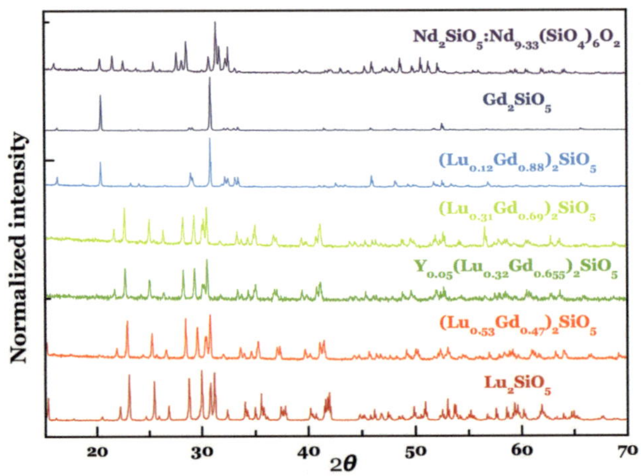

Figure 4.
XRD diffractograms of the oxyorthosilicates produced by LFZ in air at 10 mm h^{-1} [18, 45–47].

that the change from $P2_1/c$ (Gd_2SiO_5 and $(Lu_{0.12}Gd_{0.88})_2SiO_5$) to C2/c space group is most visible in the diffractograms at 2θ ranging from 20 to 30°.

Following the structural overview about oxyorthosilicates, the Raman spectroscopy analysis can be divided in two groups or families considering the total number of vibrational modes together with the shape of the high frequency region. Thus, assuming the Voron'ko et al. [8] notation, oxyorthosilicates can be distinguished among **A-type** and **B-type** silicates depending on whether a triplet or a doublet, respectively, appears inside the ($\nu_1 + \nu_3$) region [8, 9, 18, 45–47]. This way, the Raman spectra can be divided in four vibrational regions denoted as ν_3, ($\nu_1 + \nu_3$), ν_4 and $\nu_{ext} + \nu_2$ (**Figure 5**) [8]. The modes (ν_1)–(ν_4) correspond to free internal vibrations of the tetrahedral $[SiO_4]^{4-}$ complex in the reduced C_1 symmetry of the monoclinic lattice, while the external oscillations produced by the translation of the $[MO_4]$-complexes are ascribed to the ν_{ext} mode [8, 46, 52]. In addition, the modes related to rare earth ions and RE-O stretching vibrations are also placed inside the (ν_{ext} + ν_2) region.

4.1 Gadolinium oxyorthosilicate (GSO)

One of the most important oxyorthosilicates is gadolinium silicate (Gd_2SiO_5, GSO) due to the Gd^{3+} charger transfer band (CTB) that favors the transfer processes from the ground state $^8S_{7/2}$ to energy levels of the dopant element [48–50]. Transparent crystalline fibers with a yellowish aspect at naked eye (**Figure 6a**) were grown in 2017 by Rey-García et al. [18] using the LFZ technique. These fibers present a similar aspect to the ones obtained by Takagi et al. using Czochralski method [53]. Moreover, the LFZ fibers were developed in air at higher pulling rates. SEM analysis reveals a homogeneous fiber without visible grain boundaries (**Figure 6b**), thus suggesting a single crystal character. Furthermore, the EDS analysis performed confirms this homogeneity and putting in evidence the uniform elemental composition corresponding to Gd_2SiO_5 stoichiometry.

Following the structural analysis, the XRD powders diffractogram shown in **Figure 3** well matches with the 04–009-2670 XRD card (International Centre for Diffraction Data, 2019 [54]) putting in evidence the crystallinity and the monophasic nature of the monoclinic $P2_1/c$ $Gd_2(SiO_4)O$ oxyorthosilicate. Moreover, XRD scan along the longitudinal section of a polished fiber matching the diffraction maxima corresponding to {h 0 0} planes, suggesting a monocrystalline character. Aiming to confirm this evidence, 3D pole figures on longitudinal and transversal cross sections of the fibers were acquired (**Figure 7**). The crystallographic texture measurements confirmed the production of Gd_2SiO_5 single crystal fibers by LFZ, since only one high intense peak was observed in both sections. This type of morphology is potentiated by the strong thermal gradient that exists at the crystallization interface in the LFZ process [55]. So, in conclusion, these observations, namely the preferential orientation and the absence of grain boundaries in SEM analysis, permit to confirm the single crystal character of the GSO fibers grown by LFZ. In addition, the Raman spectroscopy analysis (**Figure 5**) of the GSO samples put in evidence several narrow lines, as expected for a low-symmetry crystalline structure [18].

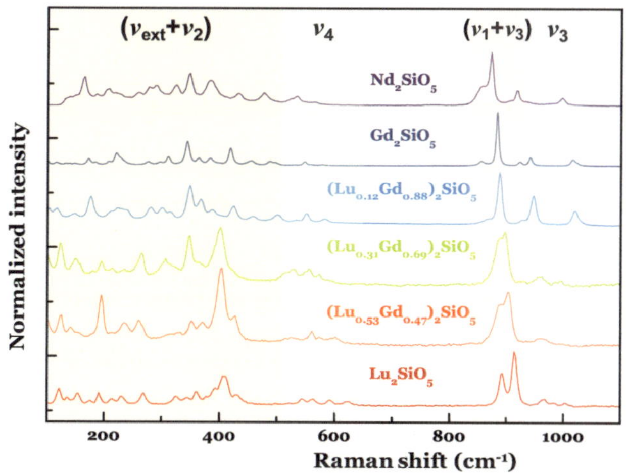

Figure 5.
Raman spectra performed under excitation of 441.6 nm line of a He-Cd laser (Kimmon IK series) of the oxyorthosilicates grown in air by LFZ [18, 45–47].

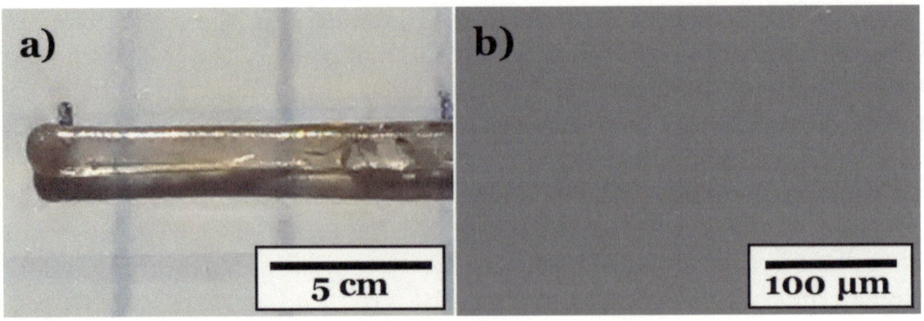

Figure 6.
(a) Photograph of GSO sample [18] and (b) corresponding SEM micrograph.

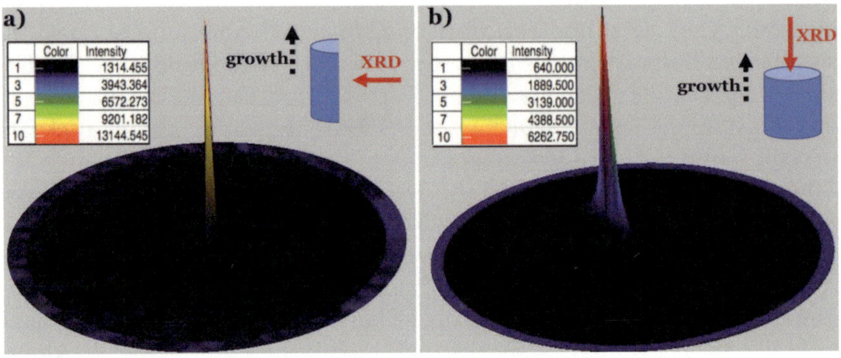

Figure 7.
XRD 3D pole figures of (a) longitudinal section, obtained for $2\theta = 30.7°$, corresponding to the (3 0 0) plane and (b) transversal cross section for (1 2 1) crystallographic plane of GSO fiber [18].

The optical spectroscopy characterization, performed from the ultraviolet to the near infrared spectral region, by photoluminescence and photoluminescence excitation suggests that GSO fibers are in fact suitable materials to be doped with rare earth active ions envisaging developing optical efficient laser materials [18]. The spectrum is mainly characterized by a series of sharp lines in the ultraviolet region corresponding to the intra $4f^7$ transitions of the Gd^{3+} ions. In fact, their excitation with ultraviolet photons promotes intra $4f^7$ transitions of the Gd^{3+} ions, and therefore, the energy transfer observed provokes internal f→f transitions of trivalent dopant ions [56–58].

4.2 Lutetium and gadolinium oxyorthosilicate (LGSO)

The interest of researchers has progressively gone *in crescendo* to mixed oxyorthosilicates of lutetium and gadolinium (LGSO) [1, 9, 59, 60] due to the following reasons:

i. Lu_2O_3 precursor is expensive, increasing the production cost of photonic materials. This way, the co-doping with gadolinium ions, using as raw material Gd_2O_3, allows reducing the cost without affecting the structural and optical properties.

ii. Gadolinium oxyorthosilicate single crystals tend to develop cracks during growth. Considering this handicap, lutetium doping has been employed aiming to reduce thermal stress and, therefore, avoiding crack formation.

iii. Finally, comparing the melting points of Lu_2O_3 (2490°C) and Gd_2O_3 (2420°C), the introduction of the second one on pure LSO materials should slightly reduce the LGSO melting point.

Analogous to what has been previously mentioned, Czochralski (CZ) method has been extensively used to develop LGSO [1, 9, 52] since Loutts et al. [61] produced LGSO for the first time in 1997. Thus, lutetium-gadolinium oxyorthosilicate crystalline fibers were successfully produced by LFZ in air at 10 mm h^{-1} [46]. Thus, three compositions based on $(Lu_xGd_{1-x})_2SiO_5$ formula were developed, establishing the doping level as x = 0.1 (LGSO-1), 0.3 (LGSO-3) and 0.5 (LGSO-5). Plane parallel-polished fragments of each one is shown in **Figure 8**. The transparency degree is clearly observed at naked eye, being not gradual with lutetium amount. In fact, LGSO-1 and LGSO-5 are translucent fibers, owning the first a yellowish tone like to that observed for pure GSO fibers. On the other hand, LGSO-3 sample is transparent, being distinguishable the colors and letters of the background image. Transmission spectra (**Figure 8**) corroborate this appearance. Transmittance values from 50% up to 77% along the visible range are observed. In addition, it must be noted that the transfer bands of the Gd^{3+} have resulted for the LGSO-3 fibers higher in intensity than pure GSO, highlighting this crystal as optimal host laser material.

On the other hand, the EDS analysis shown that LGSO crystalline fibers produced by LFZ present compositions close to the initial mixtures, in opposite to compositional dissimilarity observed on LGSO materials developed by conventional CZ method [1, 9, 52]. The expected structural change at x = 0.17 coming from the Lu amount matches with that reported by literature [1, 9, 52, 62]. Thus, LGSO-1 presents the monoclinic $P2_1/c$ structure, while the other two have a monoclinic C2/c structure, as can be deduced from the diffractograms shown in **Figure 4**. This way, LGSO-1 matched with the 01-080-9851 XRD card ICDD, while LGSO-3 and LGSO-5 are isostructural with the 00-061-0488 and the 00-061-0369 XRD cards, respectively [54].

The phase transition observed with lutetium addition is explained from atomic size and the differences of nearest surrounding rare earth ions, as reported Ryba-Romanowski et al. [63]. In fact, GSO lattice present Gd1 and Gd2 sites with different coordination number and local symmetries (CN = 9, C_{3v} and CN = 7, C_s, respectively), promoting the polyhedrons GdO_9 and GdO_7. On the other side, LSO

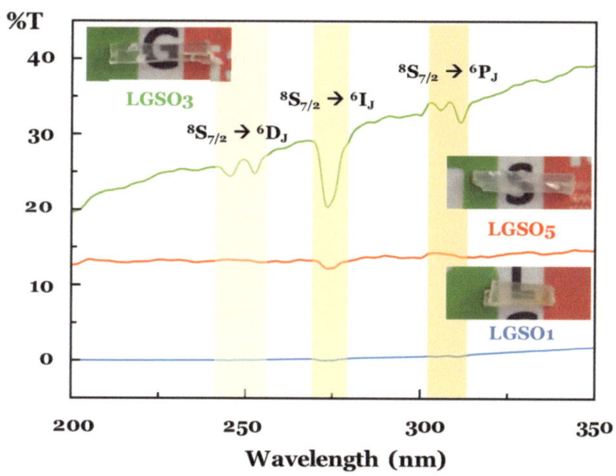

Figure 8.
Transmission spectra and photographs of LGSO fibers in the UV range [46].

lattice present both lutetium ions, and therefore, the LuO_6 (Lu1) and LuO_7 (Lu2) polyhedrons present the C_s local symmetry. The former has only one plane of symmetry, while the C_{3v} point group presents higher steric effect. Additionally, despite the smaller ionic radius of the Lu^{3+}, its inclusion into the monoclinic $P2_1/c$ unit cell strongly affects the crystalline structure. Concomitantly, the GSO structure type allows low Lu doping. In opposite, the C2/c type structure presents minor steric effect and higher symmetry degrees of freedom, allowing the introduction of large size ions such as Gd^{3+} or Ce^{3+} on the Lu sites. Consequently, LGSO-3 and LGSO-5 samples have the C2/c crystalline structure, which is larger in size than $P2_1/c$ (**Table 2**). On the other hand, **Table 2** shows how introduction of Lu decreases the volume of the cell together with an increase of the density. Concerning Raman spectroscopy characterization, LGSO-3 sample presents an intermediate spectrum between the GSO and LSO (**Figure 5**). In fact, when ($\nu_1 + \nu_3$) region is revised, LGSO-1 relates to A-type, like GSO structure, while LGSO-3 and LGSO-5 spectrum shape shows the typical doublet of B-type silicates, such as undoped LSO. Finally, the transmission studies of LGSO samples allow concluding that the Gd^{3+} ions are optically active. In fact, an intra-ionic absorption due to the energy transfer band ascribed to $^8S_{7/2} \rightarrow {}^6I_J$ transition is observed, and this band is the highest for LGSO-3 sample, the one that is the most transparent and presents lower cracks. All these considerations allow to consider the LGSO-3 fibers as the most suitable host material for photonic applications [46].

4.3 Rare-earth-doped lutetium and gadolinium oxyorthosilicates (LGSO:RE)

Once the best laser host properties were determined by developing the initial Gd_2SiO_5 to the $(Lu_{0.3}Gd_{0.7})_2SiO_5$ compositions, together with the enhancement of the structural and morphological characteristics, the scientific interest was centered in a designed doping strategy considering the most suitable rare earth ions (RE = Nd, Y and Yb).

4.3.1 Yttrium-doped (LGSO:Y)

Yttrium was selected as a dopant, since it promotes an excellent thermal and optical properties [47], being usually introduced in oxide form as a stabilizing agent. In fact, it has been largely employed in laser materials, namely yttrium aluminum garnets (YAG) or thermal barrier coatings (TBCs) due to its good thermal conductivity (13.6 W m^{-1} K^{-1}), shock resistance and low thermal expansion coefficient [64–66]. Additionally, its melting point (2425°C) is close to the one of Gd_2O_3 (2430°C).

Crystal	R_{exp} (%)	R_p (%)	R_{wp} (%)	GOF	a (Å)	b (Å)	c (Å)	Volume (10^6 pm^3)	Density (g cm^{-3})
GSO	2.82	4.23	5.70	2.08	9.128	7.058	6.746	414.26	6.775
LGSO1	2.56	2.68	3.39	1.26	9.123	7.021	6.738	411.69	6.954
LGSO3	3.33	3.75	5.29	1.59	14.461	6.750	10.495	867.06	6.570
LGSOY	2.22	3.05	4.07	1.83	14.457	6.749	10.491	866.32	6.576
LGSO5	3.29	4.23	5.51	1.68	14.391	6.716	10.425	852.72	6.949

The conventional agreement indices R_{exp}, R_p and R_{wp} correspond to the expected, profile and weighted profile R-factors, respectively. The GOF parameter represents the goodness of fit.

Table 2.
Refined unit cell parameters and relative densities of GSO and lutetium and yttrium-doped single crystals calculated from XRD analysis in powders [18, 46, 47].

The introduction of yttrium should bring stress hampering, maintaining the role of a laser passive element, by substitution of Lu^{3+} ions due to their similar atomic radius (212 pm for Y^{3+} and 217 pm for Lu^{3+}) and considering that Y_2SiO_5 has C2/c monoclinic structure. Thus, the introduction of 5 mol% of yttrium provoked a significant increase on transparency, with transmittance values around 86% along the visible range, and also reducing or even avoiding crack formation when compared with the pure LGSO (**Figure 8**). As additional advantage, the relative absorption intensity of the charge transfer bands of the Gd^{3+} ions, namely the intra $4f^7$ transitions, has been significantly increased (**Figure 9**), enhancing their suitability as matrix [47].

On the other side, the introduction of yttrium does not bring significant modifications in the lattice structure of the crystal, since XRD powder diffractogram of the LGSO:Y totally matches with the 00-061-0488 ICDD XRD card like the LGSO-3 single crystal fiber [54], and the Rietveld refinement shows similar unit cell parameters for both materials (**Table 2**). Indeed, Y^{3+} ions have substituted Gd^{3+} ions due to their close ionic radii (0.90 Å and 0.94 Å in a 6-fold coordination, respectively) along with similar electronegativity values (1.22 and 1.20, respectively) [67]. Concomitantly, Y^{3+} has increased plasticity, thus reducing stress, minimizing crack formation and maintaining the C2/c monoclinic structure.

4.3.2 Neodyminum (LGSO:Nd) and ytterbium-doped (LGSO:Yb)

It should be noted that the approach presented here is in production process.

Following the doping strategy for the $(Lu_{0.3}Gd_{0.7})_2SiO_5$ (LGSO-3) matrix, the next step was the doping with laser active elements such as Nd^{3+} and Yb^{3+} aiming to produce laser active materials. These dopants are extensively used as emitting ions in several laser materials [4–6, 33, 34] produced by different crystallization methods. Indeed, de la Fuente et al. [33] and Black et al. [34] produced $GdNdSiO_5$ and $7Gd_2O_3{\bullet}9SiO_2$:Nd single crystal laser materials by LFZ in $Ar{:}O_2$ atmospheres 30 years ago. However, most researchers employed standard growth methods. For example, Xu et al. [5] produced a controllable dual-wavelength continuous-wave laser emitting at 1075 and 1079 nm achieving an optical-to-optical efficiency of 63.3% for a Nd:Lu_2SiO_5 crystal grown by CZ, in which a peak power of 2.34 kW was measured under passively Q-switched operation. On the other side, Kim

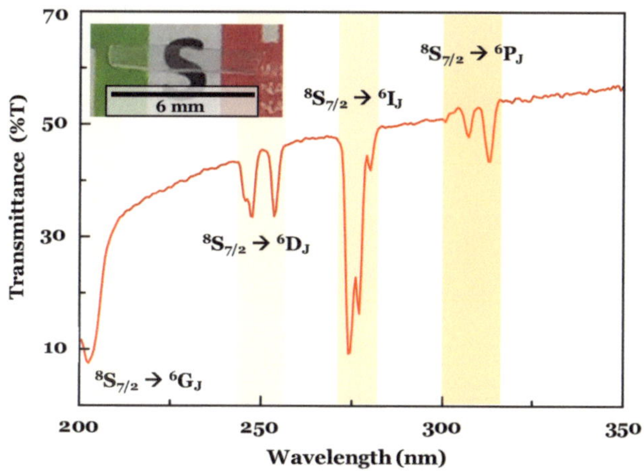

Figure 9.
Crystal photograph and transmission spectrum of LGSO:Y [47].

et al. [6] produced by hot-pressing a 10% Yb:Lu_2O_3 laser crystal pumped at 975 nm and emitting at 1080 nm presenting a slope efficiency of 74% with an output power of 16 W.

Considering the previous results of LGSO-3 samples, new powder mixtures were prepared by doping with 5 mol% of Nd_2O_3 and Yb_2O_3, using highly pure (5 N) oxide powders. Thus, neodymium (LGSO:Nd) and ytterbium (LGSO:Yb) doped LGSO single crystal fibers were produced by LFZ in air at 10 mm h^{-1}. However, these crystals present lower optical quality than those of LGSO doped with yttrium. The absorption of the LGSO:Yb crystals checked in a Z-cavity varied between 35 and 50% for a pumping of 978 nm Ti-sapphire laser, emitting at ~1039 nm wavelength (**Figure 10a**). On the other hand, LGSO:Nd diode pumped at 811 nm emits at ~1076 nm wavelength and absorbs 80% (**Figure 10b**).

4.4 Lutetium oxyorthosilicate (LSO)

Lutetium oxyorthosilicate (Lu_2SiO_5, LSO) has attracted the attention of researchers due to their favorable thermal and optical properties, which make it suitable as host materials to be used in photonics as laser media [3–6] or scintillators [7, 41, 59]. The main technique for producing LSO crystals is usually the CZ method. As an alternative, Farhi et al. [68] in 2008 grew by laser-heated pedestal growth (LHPG) rods of LSO and LSO:Ce^{3+} in air and N_2 atmosphere at 15 mm h^{-1} from square feed rods cut from a LSO pellet prepared by solid state reaction. Thus, it was expected that these type of oxyorthosilicates could be produced in air at 10 mm h^{-1} like the GSO. However, despite the fibers grown at 200 and 100 mm h^{-1} by the LFZ technique present a translucent aspect, the fibers obtained at lower pulling rates (10 and 5 mm h^{-1}) are white and opaque (insets of **Figure 11**) [45]. SEM and EDS analysis of the samples produced by LFZ put in evidence a transition from single crystal to eutectic ceramics with the gradual appearance of the Lu_2O_3 phase into the Lu_2SiO_5 matrix as pulling rate is decreased. The eutectics present a banded structure of alternated monophasic oxyorthosilicate regions with a biphasic Lu_2SiO_5/Lu_2O_3 phases. The presence of both phases was also corroborated by XRD analysis [45].

The strong difference between the melting points of both precursors, SiO_2 (1710°C) and Lu_2O_3 (2490°C), explains this behavior. In fact, the high laser power necessary to melt lutetium oxide induces SiO_2 evaporation, by overheating during laser processing. This phenomenon was already observed by Farhi et al. [68]. The most volatile compounds tend to evaporate due to overheating when two materials

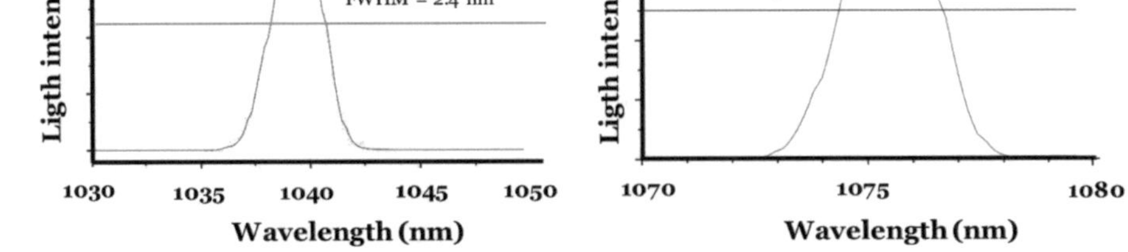

Figure 10.
Emitting wavelength of (a) LGSO:Yb and (b) LGSO:Nd crystals.

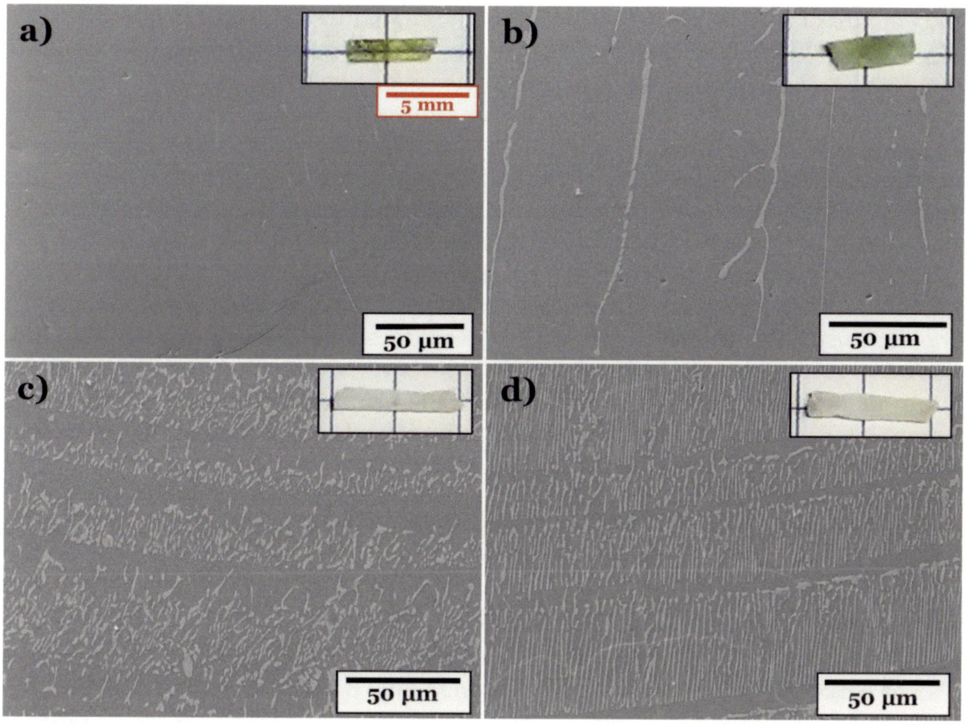

Figure 11.
SEM micrographs and photographs of LSO samples grown by LFZ at (a) 200 mm h^{-1}, (b) 100 mm h^{-1}, (c) 10 mm h^{-1} and (d) 5 mm h^{-1} [45].

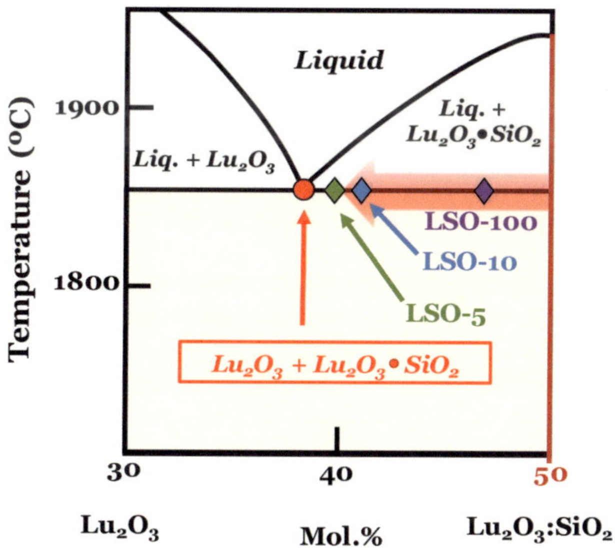

Figure 12.
Compositional deviation from the nominal 1:1 composition to eutectic point (red dot) into the Lu_2O_3-SiO_2 phase diagram based on Yb_2O_3-SiO_2 system [45, 69].

with a great melting point mismatch are processed by LFZ. Concomitantly, a deviation from the nominal composition is observed, and consequently, the as-grown material exhibits a different composition. This is what happened in the Lu_2O_3-SiO_2 system. Consequently, the crystallization path may vary, and therefore, the nature and amount of solidified phases present in the crystallized material can be different. **Figure 12** puts in evidence the compositional deviation from the nominal composition (Lu_2O_3:SiO_2 = 1:1) toward the eutectic point in the Lu_2O_3-SiO_2 binary phase diagram (red dot). This compositional shift corroborates the SEM images of **Figure 11**, where an almost complete eutectic morphology is achieved in the slower processed fibers, the ones submitted to higher power for longer times. It is noteworthy the preferential alignment of the eutectic constituent, triggered by the strong axial thermal gradients at the solidification interface [70, 71]. The introduction of other rare earth ions such as Nd, with an intermediate melting point (2233°C), contributes to the decrease of the evaporation process, since both Lu_2O_3 lamellae presence inside the Lu_2SiO_5 and their size are reduced in diameter and significantly decreased in number.

Another interesting characteristic of the LSO fibers grown by the LFZ process is a pronounced photochromic effect, from white to pink-reddish tone. This behavior was observed in samples grown at lower pulling rates when irradiated with UV light. This effect persists for samples stored in the dark, even in the presence of oxidant atmosphere, while the bleaching of the photochromic coloration is reversible when samples are under natural illumination (**Figure 13**) [45].

4.5 Neodymium oxyorthosilicate (NSO)

The Nd_2SiO_5 compound has been obtained only by solid state methods and, in the most cases, accompanied by other silicate phases, namely the pyrosilicate $Nd_2Si_2O_7$ [72–74]. In fact, it must be noted that the production of this kind of orthosilicates is hard since the Nd_2O_3-SiO_2 system melts incongruently [73]. Concomitantly, Jiang et al. [74] have recently sintered for microwave device applications, pure Nd_2SiO_5 starting from Nd_2O_3:SiO_2 = 1:1.05 mixtures. A minimum deviation from this ratio promotes the formation of by-products such as $Nd_4Si_3O_{12}$ or Nd_2O_3.

The processing of stoichiometric mixtures of SiO_2 and Nd_2O_3 oxides through the LFZ technique produced large violet fibers. The apparent crystalline aspect of these fibers increases with pulling rate from umber-like (5 and 10 mm h^{-1}) to brighter materials (100–400 mm h^{-1}) (**Figure 14a**). The XRD analysis identified the presence of two phases, which were confirmed by SEM analysis as elongated Nd_2SiO_5 crystals inside of the $Nd_{9.33}(SiO_4)_6O_2$ matrix (**Figure 14b**). The phases present a preferential orientation along the fiber axis. Furthermore, the amount of each phase significantly depends on the pulling rate. Lower pulling rates tend to increase the nonstoichiometric phase, $Nd_{9.33}(SiO_4)_6O_2$.

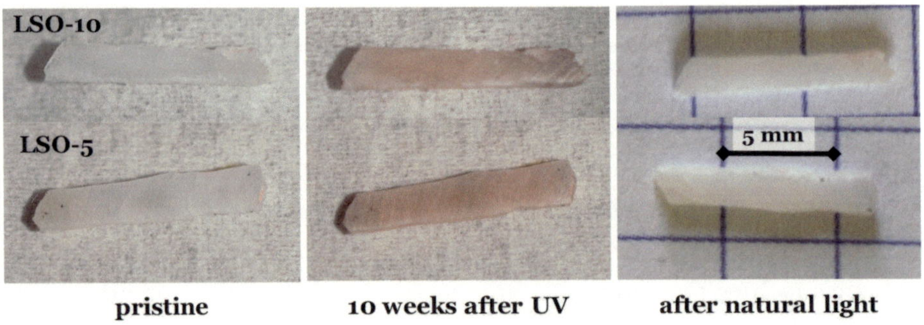

Figure 13.
Photographs of parallel-polished LSO fibers showing photochromic effect after irradiation with UV light (254 nm) [45].

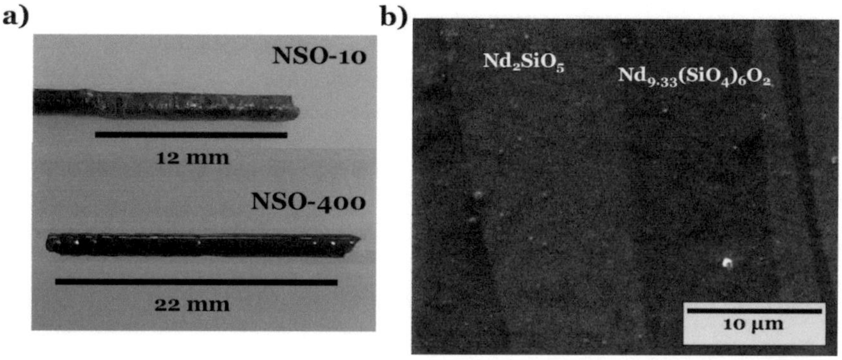

Figure 14.
(a) Photographs of NSO samples grown at 10 (NSO-10) and 400 mm h^{-1} (NSO-400), and (b) SEM micrograph identifying phases present in the NSO fiber.

Envisaging the applications of NSO fiber and considering the opposite electrical behavior of both phases together with the fibers texturing, mainly the ones grown at the highest pulling rate (200 and 400 mm h^{-1}), it is important to study their electrical properties. This way, the electrical conductivity and the dielectric constant were measured from room temperature until 1000°C and varying the frequency from 10^2 to 10^6 Hz. An increase of the AC conductivity with frequency and temperature was observed for all samples. Specifically, at 1000°C and 10 Hz, the electrical conductivity of 10^{-5} S cm^{-1} at room temperature increases to 10^{-2} S cm^{-1}. The NSO fibers exhibit a typical response of ionic conductors [30, 71] with a p-type electronic behavior, according to León-Reina et al. [75] for Nd_2SiO_5 samples. On the other side, the dielectric constant decreases with frequency while an increase with temperature is observed. Values of 10 at room temperature increase up to 10^9 at 1000°C. The high densification degree of fibers and the different polarization mechanisms, as reported Jiang et al. [74], underlie this behavior. This consideration agrees with the typical characteristics of LFZ method that is known to produce samples with a higher density when compared with standard solid state sintering [76]. Additionally, these results confirm local variations in the Nd phases leading to interstitial oxygen variations affecting the electrical response [75, 77].

5. Conclusions

This chapter puts in evidence several advantages of the LFZ technique with respect to standard growth methods. In fact, LFZ is a suitable crystallization technique that allows obtaining highly oriented refractory materials such as the rare earth oxyorthosilicates. Focusing on the special characteristics of the LFZ process and extrapolating to other hard-synthesis materials, the LFZ revealed to be a suitable method for prototyping. This consideration is based on the capabilities that directly promote a reduction of the effective production costs, namely: it is a crucible-free technique, avoiding external contamination; it allows working with low precursors amount, together with the possibility to produce low volume of high-quality materials; it is a fast processing technique that allows reducing processing time; it allows the production of small-in-size high-quality single crystal or highly textured ceramics with an appropriate geometry, allowing the development of compact and miniaturized photonic devices; it allows the growth of directionally solidified eutectic ceramic materials; it allows the crystallization of metastable phases. Thus, it was demonstrated the ability to produce high-quality single crystals of the $(Lu_{0.3}Gd_{0.7})_2SiO_5$ based composition in air at fast pulling rates. These crystals are suitable as laser passive and gain media. On the other side, Lu_2O_3/Lu_2SiO_5 eutectic ceramics exhibit an interesting reversible photochromic effect. Last, biphasic ceramics of $Nd_2SiO_5/Nd_{9.33}(SiO_4)_6O_2$ suitable for microwave devices can also be synthesized by the LFZ method.

Acknowledgements

F. Rey-García acknowledges the Portuguese Science and Technology Foundation (FCT) for the SFRH/BPD/108581/2015 grant and is grateful to funding from EU (project SPRINT, EU H2020-FET-OPEN/0426). C. Bao-Varela acknowledges funds from Xunta de Galicia (ED431E 2018/08) and Consellería de Cultura, Educación e Ordenación Universitaria (ED431B 2017/64). F.M. Costa acknowledges financial support from FEDER funds through the COMPETE 2020 Programme and National Funds through FCT—Portuguese Foundation for Science and Technology under the projects UID/CTM/50025/2019 and POCI-01-0145-FEDER-028755.

Author details

Francisco Rey-García[1*], Carmen Bao-Varela[2] and Florinda M. Costa[3]

1 Instituto de Ciencia de Materiales de Aragón (ICMA-CSIC), Zaragoza, Spain

2 UA Microoptics and GRIN Optics group (ICMA-CSIC), Universidade de Santiago de Compostela, Santiago de Compostela, Spain

3 i3N and Physics Department, Universidade de Aveiro, Aveiro, Portugal

*Address all correspondence to: francisco.rey.usc@gmail.com

References

[1] Ryba-Romanowski SA, Lisiecki R, Berkowski M, Rodriguez-Rodriguez H, Martin IR. Effect of substitution of lutetium by gadolinium on emission characteristics of $(Lu_xGd_{1-x})_2SiO_5$:Sm^{3+} single crystals. Optical Materials Express. 2014;**4**:739-752. DOI: 10.1364/OME.4.000739

[2] Kobayashi K, Sakka Y. Research progress in nondoped lanthanoid silicate oxyapatites as new oxygen-ion conductors. Journal of the Ceramic Society of Japan. 2004;**122**:921-939. DOI: 10.2109/jcersj2.122.921

[3] Zheng L, Su L, Growth XJ. Characterization of ytterbium doped silicate crystals for ultra-fast laser applications. In: Kolesnikov N, Borisenko E, editors. Modern Aspects of Bulk Crystal and Thin Film Preparation. Rijeka: IntechOpen; 2012. pp. 25-42. DOI: 10.5772/30457

[4] Li DZ, Xu XD, Zhou DH, Xia CT, Wu F, Xu J, et al. Crystal growth, optical properties, and continuous-wave laser operation of Nd^{3+}-doped Lu_2SiO_5 crystal. Laser Physics Letters. 2011;**8**: 32-37. DOI: 10.1002/lapl.201010092

[5] Xu X, Di J, Zhang J, Tang D, Xu J. CW and passively Q-switched laser performance of Nd:Lu_2SiO_5 crystal. Optical Materials. 2016;**51**:241-244. DOI: 10.1016/j.optmat.2015.11.008

[6] Kim W, Villalobos G, Baker C, Frantz J, Shaw B, Bayya S, et al. Overview of transparent optical ceramics for high-energy lasers at NRL. Applied Optics. 2015;**54**:F210-F221. DOI: 10.1364/AO.54.00F210

[7] Jary V, Krasnikov A, Nikl M, Zazubovich S. Origin of slow low-temperature luminescence in undoped and Ce-doped Y_2SiO_5 and Lu_2SiO_5 single crystals. Physica Status Solidi B: Basic Solid State Physics. 2015;**252**:274-281. DOI: 10.1002/pssb.201451234

[8] Voron'ko YK, Sobol AA, Shukshin VE, Zagumennyi AI, Zavartsev YD, Koutovoi SA. Spontaneous Raman spectra of the crystalline, molten and vitreous rare-earth oxyorthosilicates. Optical Materials. 2011;**33**:1331-1337. DOI: 10.1016/j.optmat.2011.03.021

[9] Bińczyk M, Glowacki M, Lapiński A, Berkowski M, Runka T. μ-Raman and infrared reflectance spectroscopy characterization of $(Lu_{1-x}Gd_x)_2SiO_5$ solid solution single crystals doped with Dy^{3+} or Sm^{3+}. Journal of Molecular Structure. 2016;**1109**:50-57. DOI: 10.1016/j.molstruc.2015.12.078

[10] Kamada K, Murakami R, Kochurikhin VV, Luidmila G, Kim KJ, Shoji Y, et al. Single crystal growth of submillimeter diameter sapphire tube by the micro-pulling down method. Journal of Crystal Growth. 2018;**492**:45-49. DOI: 10.1016/j.jcrysgro.2018.03.023

[11] Masubuchi Y, Higuchi M, Katase H, Takeda T, Kikkawa S, Kodaira K, et al. Oxide ion conduction in $Nd_{9.33}(SiO_4)_6O_2$ and $Sr_2Nd_8(SiO_4)_6O_2$ single crystals grown by floating zone method. Solid State Ionics. 2004;**166**:213-217. DOI: 10.1016/j.ssi.2003.09.019

[12] Liu QY, Liu ZY, Zhou XB, Liu ZG, Huo MX, Wang XJ, et al. Large size single crystal growth of Ti_4O_7 by the floating-zone method. Crystal Growth & Design. 2019;**19**:730-736. DOI: 10.1021/acs.cgd.8b01318

[13] Carvalho RG, Pires MS, Fernandes AJS, Silva RF, Costa FM. Directionally solidified eutectic and off-eutectic mullite-zirconia fibres. Journal of the European Ceramic Society. 2013;**33**:953-963. DOI: 10.1016/j.jeurceramsoc.2012.09.032

[14] Andreeta MRB, Hernandes AC. Laser-heated pedestal Growth of oxide Fibers. In: Dhanaraj G, Byrappa K, Prasad V, Dudley M, editors. Springer Handbook of Crystal Growth. Berlin: Springer; 2010. pp. 393-432. DOI: 10.1007/978-3-540-74761-1_13

[15] Soares MRN, Soares MJ, Fernandes AJS, Rino L, Costa FM, Monteiro T. YSZ: Dy^{3+} single crystal white emitter. Journal of Materials Chemistry. 2011;**21**:15262-15265. DOI: 10.1039/C1JM12564H

[16] Carvalho RG, Fernandes AJS, Oliveira FJ, Alves E, Franco N, Louro C, et al. Single and polycrystalline mullite fibres grown by laser floating zone technique. Journal of the European Ceramic Society. 2010;**30**:3311-3318. DOI: 10.1016/j.jeurceramsoc.2010.07.033

[17] Soares MRN, Nico C, Peres M, Ferreira N, Fernandes AJS, Monteiro T, et al. Structural and optical properties of europium doped zirconia single crystals fibers grown by laser floating zone technique. Journal of Applied Physics. 2011;**109**:013516. DOI: 10.1063/1.3527914

[18] Rey-García F, Ben Sedrine N, Soares MR, Fernandes AJS, Lopes AB, Ferreira NM, et al. Structural and optical characterization of Gd_2SiO_5 crystalline fibres obtained by laser floating zone. Optical Materials Express. 2017;**7**:868-879. DOI: 10.1364/OME.7.000868

[19] Costa FM, Silva RF, Vieira JM. Diffusion phenomena and crystallization path during the growth of LFZ Bi-Sr-Ca-Cu-O superconducting fibres. Superconductor Science and Technology. 2001;**14**:910-920. DOI: 10.1088/0953-2048/14/11/305

[20] Edmonds WR. The reflaxicon, a new reflective optical element and some applications. Applied Optics. 1973;**12**:1940-1944. DOI: 10.1364/AO.12.001940

[21] Martin CW. Reflecting Optical Objective System. U.S. Patent 2457253. 1948.

[22] Ferreira NM, Kovalevsky AV, Costa FM, Frade JR. Processing effects on properties of $(Fe,Mg,Al)_3O_4$ spinels as potential consumable anodes for pyroelectrolysis. Journal of the American Ceramic Society. 2016;**99**:1-5. DOI: 10.1111/jace.14190

[23] Carrasco MF, Silva RF, Vieira JM, Costa FM. Electrical field freezing effect on laser floating zone (LFZ)-grown $Bi_2Sr_2Ca_2Cu_4O_{11}$ superconducting fibres. Superconductor Science and Technology. 2004;**17**:612-619. DOI: 10.1088/0953-2048/17/4/008

[24] Carrasco MF, Silva RF, Vieira JM, Costa FM. Pulling rate and current intensity competition in electrically assisted laser floating zone. Superconductor Science and Technology. 2009;**22**:065016. DOI: 10.1088/0953-2048/22/6/065016

[25] Ferreira NM, Rasekh S, Costa FM, Madre MA, Sotelo A, Diez JC, et al. New method to improve the grain alignment and performance of thermoelectric ceramics. Materials Letters. 2012;**83**:144-147. DOI: 10.1016/j.matlet.2012.05.131

[26] Costa FM, Ferreira NM, Rasekh S, Fernandes AJS, Torres MA, Madre MA, et al. Very large superconducting currents induced by growth tailoring. Crystal Growth & Design. 2015;**15**:2094-2101. DOI: 10.1021/cg5015972

[27] Graça MPF, Peixoto MV, Ferreira N, Rodrigues J, Nico C, Costa FM, et al. Optical and dielectric behaviour of $EuNbO_4$ crystals. Journal of Materials Chemistry C. 2013;**1**:2913-2919. DOI: 10.1039/C3TC00793F

[28] Soares MRN, Nico C, Rodrigues J, Peres M, Soares MJ, Fernandes AJS, et al. Bright room-temperature green luminescence from YSZ:Tb^{3+}. Materials Letters. 2011;**65**:1979-1981. DOI: 10.1016/j.matlet.2011.03.099

[29] Rasekh S, Costa FM, Ferreira NM, Torres MA, Madre MA, Diez JC, et al. Use of laser technology to produce high thermoelectric performances in $Bi_2Sr_2Co_{1.8}O_x$. Materials and Design. 2015;**75**:143-148. DOI: 10.1016/j.matdes.2015.03.005

[30] Carvalho RG, Fernandes AJS, Silva RF, Costa FM, Figueiredo FM. Directional solidification of ZrO_2-$BaZrO_3$ composites with mixed protonic-oxide ionic conductivity. Solid State Ionics. 2014;**262**:654-658. DOI: 10.1016/j.ssi.2013.10.051

[31] Kobayashi K, Sakka Y. Rudimental research progress of rare-earth silicate oxyapatites: Their identification as a new compound until discovery of their oxygen ion conductivity. Journal of the Ceramic Society of Japan. 2014;**122**:649-663. DOI: 10.2109/jcersj2.122.649

[32] Toropov NA, Bondar IA. Lanthanum silicate $2La_2O_3$•$3SiO_3$. Bulletin of the Academy of Sciences of the USSR, Division of Chemical Science. 1959;**8**:528-530

[33] De La Fuente GF, Black LR, Andrauskas DM, Verdún HR. Growth of Nd-doped rare earth silicates by the laser floating zone method. Solid State Ionics. 1989;**32-33**:494-505. DOI: 10.1016/0167-2738(89)90261-0

[34] Black LR, Andrauskas DM, de la Fuente GF, Verdún HR. Laser heated pedestal growth of Nd-doped oxide crystals for diode pumping. In: Proceedings SPIE 1104, Growth, Characterization, and Applications of Laser Host and Nonlinear Crystals. 1989. p. 175. DOI: 10.1117/12.960593

[35] Duan X, Qian C, Shen Y, Su L, Zheng L, Li L, et al. Efficient Ho:$(Sc_{0.5}Y_{0.5})_2SiO_5$ laser at 2.1 μm in-band pumped by Tm fiber laser. Optics Express. 2019;**27**:4522-4527. DOI: 10.1364/OE.27.004522

[36] Seminko V, Maksimchuk P, Bespalova I, Malyukin Y. Different roles of Ce^{3+} optical centers in oxyorthosilicate nanocrystals at X-ray and UV excitation. Crystals. 2019;**9**:114. DOI: 10.3390/cryst9020114

[37] Ren X, Tian Z, Zhang J, Wang J. Equiatomic quaternary $(Y_{1/4}Ho_{1/4}Er_{1/4}Yb_{1/4})_2SiO_5$ silicate: A perspective multifunctional thermal and environmental barrier coating material. Scripta Materialia. 2019;**168**:47-50. DOI: 10.1016/j.scriptamat.2019.04.018

[38] Zhou PQ, Wang XJ, He YD, Wu ZF, Du JL, Fu EG. Effect of deposition mechanisms on the infrared photoluminescence of erbium-ytterbium silicate films under different sputtering methods. Journal of Applied Physics. 2019;**125**:175114. DOI: 10.1063/1.5089677

[39] Hopkins RH, Rowland GW, Steinbruege KB, Partlow WD. Silicate oxyapatites: New high-energy storage laser host for Nd^{3+}. Journal of the Electrochemical Society. 1971;**118**: 637-639. DOI: 10.1149/1.2408128

[40] Wallenberger FT. Commercial and experimental glass fibers. In: Wallenberger FT, Bingham PA, editors. Fiberglass and Glass Technology. Boston, MA: Springer; 2010. pp. 3-90. DOI: 10.1007/978-1-4419-0736-3_1

[41] Wu Y, Koschan M, Foster C, Melcher CL. Czochralski growth, optical, scintillation, and defect properties of Cu^{2+} codoped Lu_2SiO_5:Ce^{3+} single crystals. Crystal Growth & Design.

2019;**19**:4081-4089. DOI: 10.1021/acs.cgd.9b00479

[42] Haile HT, Dejene FB. Effect of substrate temperature on the material properties of the Y_2SiO_5: Ce^{3+} thin film by pulsed laser deposition (PLD) method. Optik. 2019;**184**:508-517. DOI: 10.1016/j.ijleo.2019.05.003

[43] Jiang F, Cheng L, Wei H, Wang Y. Hot corrosion behavior of Lu_2SiO_5 and La_2SiO_5 in a molten Na_2SO_4 environment: A first-principles corrosion resistance investigation. Ceramics International. 2019;**45**:15532-15537. DOI: 10.1016/j.ceramint.2019.05.058

[44] Nieai AA, Mohammadi M, Bahaabad MS, Arefi AH. Synthesis and characterization of spherical Yb_2SiO_5 powder using solid-state diffusional reaction and spray dry process. Journal of the Australian Ceramic Society. 2019:1-8. DOI: 10.1007/s41779-019-00353-3

[45] Rey-García F, Ben Sedrine N, Fernandes AJS, Monteiro T, Costa FM. Shifting Lu_2SiO_5 crystal to eutectic structure by laser floating zone. Journal of the European Ceramic Society. 2018;**38**:2059-2067. DOI: 10.1016/j.jeurceramsoc.2017.11.003

[46] Rey-García F, Fernandes AJS, Costa FM. Influence of Lu content on $(Lu_xGd_{1-x})_2SiO_5$ oxyorthosilicates grown by laser floating zone: Structural studies and transparency. Materials Research Bulletin. 2018;**112**:413-419. DOI: 10.1016/j.materresbull.2018.04.058

[47] Rey-García F, Rodrigues J, Fernandes AJS, Soares MR, Monteiro T, Costa FM. $(Lu_{0.3}Gd_{0.7})_2SiO_5$:$Y^{3+}$ single crystals grown by the laser floating zone method: Structural and optical studies. CrystEngComm. 2018;**20**:7386-7394. DOI: 10.1039/C8CE01319E

[48] Ananias D, Kostova M, Almeida Paz FA, Ferreira A, Carlos LD, Klinowski J, et al. Photoluminescent layered lanthanide silicates. Journal of the American Chemical Society. 2004;**126**:10410-10417. DOI: 10.1021/ja047905n

[49] Sá Ferreira RA, Karmaoui M, Nobre S, Carlos LD, Pinna N. Optical properties of lanthanide-doped lamellar nanohybrids. ChemPhysChem. 2006;**7**:2215-2222. DOI: 10.1002/cphc.200600317

[50] Hernández-Adame L, Méndez-Blas A, Ruiz-García J, Vega-Acosta JR, Medellín-Rodríguez FJ, Palestino G. Synthesis, characterization, and photoluminescence properties of Gd:Tb oxysulfide colloidal particles. Chemical Engineering Journal. 2014;**258**:136-145. DOI: 10.1016/j.cej.2014.07.067

[51] Springer Materials [Internet]. 2019. Available from: https://materials.springer.com [Accessed: 01 August 2019]

[52] Dominiak-Dzik G, Ryba-Romanowski W, Lisiecki R, Solarz P, Macalik B, Berkowski M, et al. The Czochralski growth of $(Lu_{1-x}Gd_x)_2SiO_5$:Dy single crystals: Structural, optical, and dielectric characterization. Crystal Growth & Design. 2010;**10**:3522-3530. DOI: 10.1021/cg100429b

[53] Takagi K, Fukuzawa T. Cerium-activated Gd_2SiO_5 single crystal scintillator. Applied Physics Letters. 1983;**42**:43-45. DOI: 10.1063/1.93760

[54] International Centre for Diffraction Data [Internet]. 2019. Available from: http://www.icdd.com [Accessed: 01 August 2019]

[55] Vieira JM, Silva RA, Silva RF, Costa FM. Enhancement of superconductivity in LFZ-grown BSCCO fibres by steeper axial temperature gradients. Applied Surface

Science. 2012;**258**:9175-9180. DOI: 10.1016/j.apsusc.2011.11.054

[56] Blasse G, Grabmaier BC. Luminescent Materials. Berlin: Wiley; 1994. DOI: 10.1007/978-3-642-79017-1

[57] Lushchik A, Lushchik C, Nagirnyi V, Pazylbek S, Sidletskiy O, Schwartz K, et al. On the mechanism of radiation damage and prospects of their suppression in complex metal oxides. Physica Status Solidi B: Basic Solid State Physics. 2013;**250**:261-270. DOI: 10.1002/pssb.201200488

[58] Lushchik A, Nagirnyi V, Shablonin E, Sidletskiy O, Toxanbayev B, Zhunusbekov A. Luminescence of Cations Excitons in Gd_2SiO_5 Crystals [Internet]. Available from: http://photon-science.desy.de/annual_report/files/2009/2009545.pdf [Accessed: 01 August 2017]

[59] Kervalishvili J, Yannakopoulos PH. Nuclear Radiation Nanosensors and Nanosensory Systems. Dordrecht: Springer; 2016. DOI: 10.1007/978-94-017-7468

[60] Lee S, Lee MS, Won JY, Lee JS. Performance of a new accelerating-electrode-equipped fast-time-response PMT coupled with fast LGSO. Physics in Medicine and Biology. 2018;**63**:05NT03. DOI: 10.1088/1361-6560/aaad20

[61] Loutts GB, Zagumennyi AI, Lavrishchev SV, Zavartsev YD, Studenikin PA. Czochralski growth and characterization of $(Lu_{1-x}Gd_x)_2SiO_5$ single crystals for scintillators. Journal of Crystal Growth. 1997;**174**:331-336. DOI: 10.1016/S0022-0248(96)01171-2

[62] Zorenko Y, Gorbenko V, Savchyn V, Zorenko T, Grinyov B, Sidletskiy O, et al. Growth and luminescent properties of Ce and Ce-Tb doped $(Y,Lu,Gd)_2SiO_5$:Ce single crystalline films. Journal of Crystal Growth. 2014;**401**:577-583. DOI: 10.1016/j.jcrysgro.2014.01.066

[63] Ryba-Romanowski W, Macalik B, Berkowski M. Down- and up-conversion of femtosecond light pulse excitation into visible luminescence in cerium-doped Lu_2SiO_5-Gd_2SiO_5 solid solution crystals co-doped with Sm^{3+} or Dy^{3+}. Optics Express. 2015;**23**:4552-4562. DOI: 10.1364/OE.23.004552

[64] Kumar R, Jordan E, Gell M, Roth J, Jiang C, Wang J, et al. CMAS behavior of yttrium aluminum garnet (YAG) and yttria-stabilized zirconia (YSZ) thermal barrier coatings. Surface and Coating Technology. 2017;**327**:126-138. DOI: 10.1016/j.surfcoat.2017.08.023

[65] Guarino S, Ponticelli GS, Giannini O, Genna S, Trovalusci F. Laser milling of yttria-stabilized zirconia by using a Q-switched Yb:YAG fiber laser: Experimental analysis. International Journal of Advanced Manufacturing Technology. 2018;**94**:1373-1385. DOI: 10.1007/s00170-017-1020-8

[66] Mun JH, Jouini A, Yoshikawa A, Kim JH, Fukuda T, Lee JS. Thermal and optical properties of Yb-doped Lu_2O_3 single crystal grown by the micro-pulling-down method. Journal of Ceramic Processing Research. 2013;**14**:636-640. Available from: http://jcpr.kbs-lab.co.kr/file/JCPR_vol.14_2013/JCPR14-5/10.pdf [Accessed: 01 August 2017]

[67] Zorenko Y, Gorbenko V, Savchyn V, Voznyak T, Grinyov B, Sidletskiy O, et al. Growth and luminescent properties of Lu_2SiO_5:Ce and $(Lu_{1-x}Gd_x)_2SiO_5$:Ce single crystalline films. Journal of Crystal Growth. 2011;**337**:72-80. DOI: 10.1016/j.jcrysgro.2011.10.003

[68] Farhi H, Lebbou K, Belkahla S, Grosvalet L, Hautefeuille B, Caramanian A, et al. Fiber single crystal

growth by LHPG technique and optical characterization of Ce^{3+}-doped Lu_2SiO_5. Optical Materials. 2008;**30**:1461-1467. DOI: 10.1016/j.optmat.2007.09.002

[69] Peak JD, Melcher CL, Rack PD. Combinatorial thin film synthesis of cerium doped scintillation materials in the lutetium oxide-silicon oxide system. IEEE Transactions on Nuclear Science. 2008;**55**:1480-1483. DOI: 10.1109/TNS.2008.922839

[70] Carvalho RG, Oliveira FJ, Silva RF, Costa FM. Mechanical behaviour of zirconia-mullite directionally solidified eutectics. Materials and Design. 2014;**61**:211-216. DOI: 10.1016/j.matdes.2014.04.050

[71] Carvalho RG, Kovalevsky AV, Lufaso MW, Silva RF, Costa FM, Figueiredo FM. Ionic conductivity of directionally solidified zirconia-mullite eutectics. Solid State Ionics. 2014;**256**: 45-51. DOI: 10.1016/j.ssi.2013.12.033

[72] Masubuchi Y, Higuchi M, Kodaira K. Reinvestigation of phase relations around the oxyapatite phase in the Nd_2O_3•SiO_2 system. Journal of Crystal Growth. 2003;**247**:207-212. DOI: 10.1016/S0022-0248(02)01908-5

[73] Saal JE. Thermodynamic modelling of the reactive sintering of Nd:YAG [thesis]. Pennsylvania (PA) USA: Pennsylvania State University; 2008

[74] Jiang C, Wu S, Ma Q, Mei Y. Synthesis and microwave dielectric properties of Nd_2SiO_5 ceramics. Journal of Alloys and Compounds. 2012;**544**:141-144. DOI: 10.1016/j.jallcom.2012.07.076

[75] León-Reina L, Porras-Vázquez JM, Losilla ER, Moreno-Leal L, Aranda MAG. Structure and oxide anion conductivity in $Ln_2(TO_4)O$ (*Ln*= La, Nd; T= Ge, Si). Journal of Solid State Chemistry. 2008;**181**:2501-2506. DOI: 10.1016/j.jssc.2008.06.029

[76] Ferreira NM, Kovalevsky AV, Costa FM, Frade JR. Processing effects on properties of $(Fe,Mg,Al)_3O_4$ Spinels as potential consumable anodes for pyroelectrolysis. Journal of the American Ceramic Society. 2016;**99**:1889-1893. DOI: 10.1111/jace.14190

[77] Nakayama S, Sakamoto M, Higuchi M, Kodaira K, Sato M, Kakita S, et al. Oxide ionic conductivity of apatite type $Nd_{9.33}(SiO_4)_6O_2$ single crystal. Journal of the European Ceramic Society. 1999;**19**:507-510. DOI: 10.1016/S0955-2219(98)00215-5

Chapter 7

Solid Phase Evolution of Nanodispersed Palladium Powders

Veronika Ivanovna Rozhdestvina

Abstract

The processes of solid phase evolution of nanodispersed palladium powder at low temperatures were studied. It has been established that the process of solid phase transformation, which develops over time, forms a hierarchically structured organization of palladium grains from a structurally loose atomic cluster to a micrograin—an encapsulated aggregate of hollow subgrains. The process of grain ordering unfolds at several scale levels. It starts with the inner walls of the hollow subgrains that form the channel structures of the microaggregate and then passes to their surface and the unified encapsulating grain shell. In the collective effects of self-organization, periodic activation of mass transfer is observed, in which nanoparticles of various mesoscale structure organization are involved.

Keywords: palladium, nanodispersed powders, solid phase evolution

1. Introduction

Palladium is one of the most versatile elements of the platinum group metals. Ultrafine palladium particles exhibit unique catalytic properties (extremely high activity and selectivity); they are used to create composite materials and are capable of absorbing large amounts of hydrogen [1]. Current trends in the study of nanostructured palladium are mainly aimed at the development of new methods for the manufacture and stabilization of nanoparticles [1–5] and control of sizes and shapes that determine the physicochemical properties [6]. Despite numerous theoretical and experimental studies, structural-reactional relationships and the understanding of mechanisms of the catalytic reaction on the surface of palladium nanoparticles remain controversial.

In the study of evolutionary interactions between nanoparticles of the solid phase, the emphasis is on the mass exchange theory, experimental and model studies of agglomeration in a dispersion medium (liquid or gaseous). Contact reactions occur when particles approach each other randomly or in a controlled manner [7–10]. Most studies of the contact interactions of metallic dispersed powders are performed on systems in highly excited states caused by deformation influences of thermally activated nature [11, 12]. As a result, processes go with high velocities, causing the contact melting effect—sintering between contacting surfaces.

However, very little attention is paid to the study of the subsequent solid phase evolution of a nanostructured system under conditions of surface energy dissipation, without deformation and thermal effects, since it is believed that solid phase processes are not large-scale and proceed with extreme slowness. In this regard, a study of structural and morphological transformations of nanodispersed palladium powders at low temperatures is of both practical importance and fundamental scientific interest.

In the framework of this article, we studied the processes and mechanisms of solid phase transformations that occur in freely contacting nanodispersed palladium powders at low temperatures.

2. Materials and methods of study

We studied PdAP-0 nanodispersed powders manufactured by Krastsvetmet JSC (mass fraction of Pd min. 99.98%) obtained by chemical reduction of palladium from solution, followed by filtration, washing and drying, packed in glass ampoules (1 month from the date of manufacture to the beginning of the experiment). The powder was agitated in ethanol, and a drop of suspension was applied to a conductive carbon tape fixed to observation tables in an electron microscope. After the droplet dried, the samples were placed in a vacuum column of an electron microscope.

Among all elements of the platinum group, the Debye characteristic temperature Pd Θ_D = 274 K is closest to the temperature of normal conditions. To observe changes in the morphostructural characteristics of a nanodispersed powder, the particles of which are in free contact; the following temperature regimes were chosen for experimental observations of solid phase transformation. The first temperature regime which is 293 ± 5 K exceeds Θ_D Pd by no more than 24°, that is, the increase in the vibration amplitude due to the temperature effects is insignificant. The second temperature regime is 258 ± 1 K, which is lower than Θ_D Pd by 16°. The amplitude of the natural vibrations of atoms is still significant under these conditions, but the temperature contribution to them is minimal. The third temperature regime is 77 K, which is lower than Θ_D Pd by 197°. The amplitude of natural vibrations of atoms is low, and interatomic interactions dominate.

The powder was placed in tightly closed ampoules, from which air was evacuated to prevent moisture condensation on the samples when operating at low temperatures. The one ampoule was kept at room temperature, the second in a cryostat (258 ± 1 K) and the third in liquid nitrogen (77 K). The total duration of the experiment was 2 years. Control observations of evolutionary changes in the morphostructural features of palladium powders in ampoules were carried out after every 6 months of exposure of the samples to the respective conditions. After this period, the ampoules were removed and stabilized without opening to room temperature during the day. Powder samples were examined with an optical microscope for the occurrence of larger particles in the powder. If available, individual particles were extracted from the powder, and studies of their structural and morphological characteristics were carried out. Then the particles were again placed in ampoules from which air was evacuated, and the ampoules were again placed in the appropriate temperature conditions for the next holding period.

Systematic observations of solid phase transformation processes in finely dispersed palladium powders were performed using the device base of the Analytical Center of Mineralogical and Geochemical Research of the Institute of Geology and Nature Management, Far-Eastern Branch of the Russian Academy of Sciences: JEOL JSM 6390LV (Japan) scanning electron microscope and SIGMA (Carl Zeiss) scanning electron microscope with the X-Max INCA Energy (Oxford Instrument). X-ray diffraction studies were performed using a Shimadzu XRD-7000 MAXima. X-ray diffractometer (2.2 kW, Cu Target, Long Fine Focus (LFF) type with a CM-3121 diffracted beam monochromator) with an MDA-1101 (with Microscope unit) attachment for analyzing microobjects (a locality region of 1.00 mm), Bruker AXS Discover D8 X-ray microdiffractometer on CuK_α radiation, with rotation, vibrations and by points (Collective Use Centre of the Far Eastern Geological Institute of the Far Eastern Branch of the Russian Academy of Sciences) and photographic method on URS-2 X-ray unit (CuK_α radiation with Ni filter) with rotating and without rotating the sample using the RKD-57.3 camera. Lattice parameters were corrected by the least-squares method using all reflections available for measurements.

3. Solid phase processes of self-organization of substance at nanolevel

In the chemical reduction of palladium from a solution at an early stage of nucleation, metal ions are reduced to a state with zero valency and, when approaching, are combined into atomic clusters to form irreversible nuclei (0.1–1 nm in size) [1]. These nuclei can increase in size due to the addition of other atoms or coagulation among them, forming floccules. To reduce the effect of coagulation and stabilization of nanoparticles, special agents are introduced into solutions that provide a monodisperse state and the required forms of nanoparticles [5, 6].

Thus, metal particles (dispersed phase) are released from the solution (dispersion medium), which are suspended in a liquid. The interactions between them lead to clustering (the first stage of substance integration). Clusters of various shapes gradually increase in size and form loose flocs. After liquid removal, the system also remains biphasic: solid particles in the form of floccules with a developed surface and gas (air) filling the spaces between them. The system is energy saturated, and it is in a non-equilibrium state. Metallic materials are dissipative systems capable of energy dissipation. This can be manifested in the activation of morphostructural solid phase transformations.

An analysis of the Debye powder diagrams obtained from the starting powders confirms their ultrafine state. The X-ray diffraction pattern of palladium powder is characterized by broadening of symmetric diffraction peaks for all crystallographic directions, with an increase in the degree of blurriness and intensity with an increase in the diffraction angle of reflection (**Figure 1**). Doublet (422) is not split. The diffraction broadening of reflections caused by a decrease in coherent dissipation blocks begins at crystallite sizes less than 100 nm. In the area of small angles, a wide halo is observed, which is characteristic of X-ray amorphous phases, due to the presence of a significant number of particles with sizes less than 10 nm. To study the degree of broadening of diffraction reflections, we used a method based on extracting from a graphical representation of the dependence of the intensity on the wavelength of a linear optical spectrum obtained by scanning X-ray diffraction patterns and then storing it in the form of a full-profile bi-dimensional description. It was found that the diffraction peaks have a symmetrical shape and the broadening is proportional to the tangent of the diffraction reflection angle, which indicates the dispersion of the powder under study. The lines with Miller indices (hkl) $h = k$ are approximated by Gaussian curves, which also indicates a high dispersion. An analysis of the degree of broadening of the diffraction reflections of the X-ray diffraction pattern of the initial palladium powder indicates that according to the size criterion of the composing particles, the powder can be divided into two dominant fractions of 30–60 nm and less than 10 nm.

Electron microscopic studies show that palladium particles are a quasi-amorphous substance in which isolated particles of two scale levels are identified: 5–20 nm and 40–150 nm (**Figure 2**). Among the first group, rounded particles with

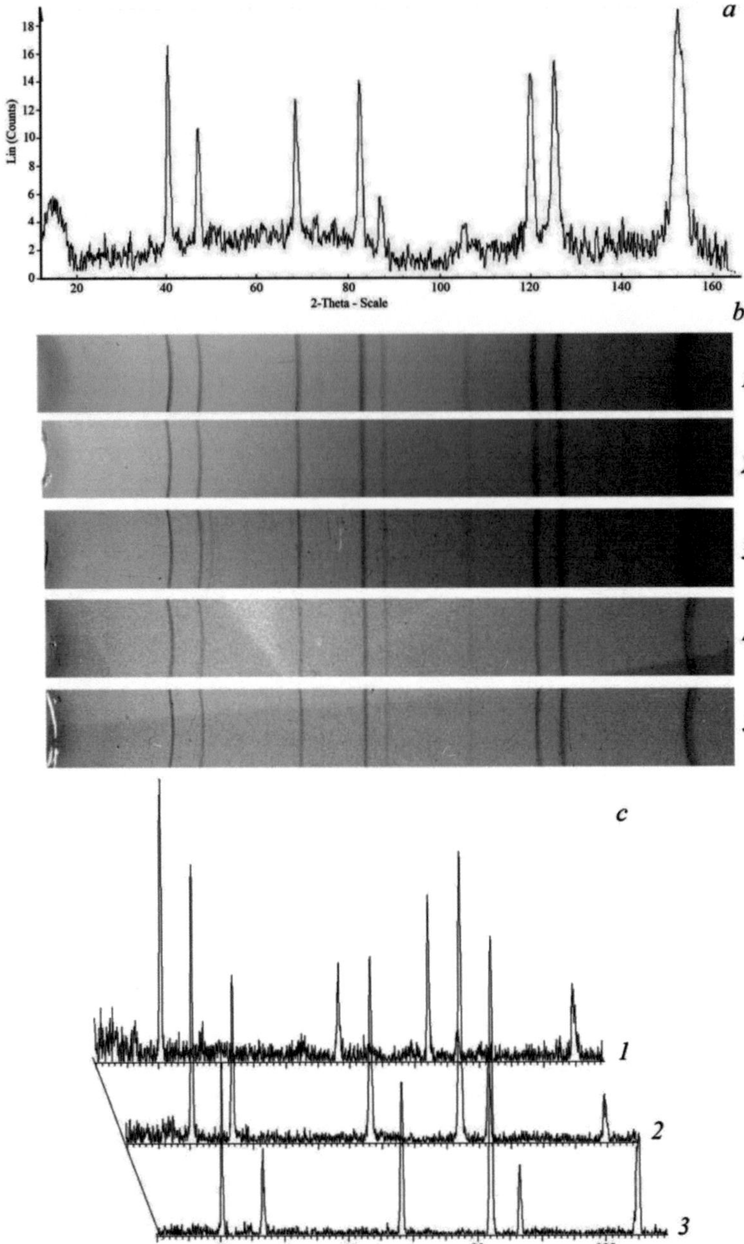

Figure 1.
X-ray diffraction patterns of palladium powder (T = 293 K)—initial state (a, b1) and state after 6 (b2), 12 (b3), 18 (b4) and 24 (b5) months. Changes in the preferred orientations of crystallites in palladium grains at different stages of conversion (state after 6 (c1), 12 (c2), 24 (c3) months).

fuzzy boundaries of 5.5–7 nm in size predominate; those larger than 15–20 nm are observed occasionally. The second group is mainly represented by particles of various configurations formed as a result of aggregation of particles of the first group or having a quasi-amorphous structure without isolation of individual elements with a predominant size of ~60 nm (**Figure 2**). At this stage, differentiation is already observed in the total mass of particles with the individualization of elements of a denser structure.

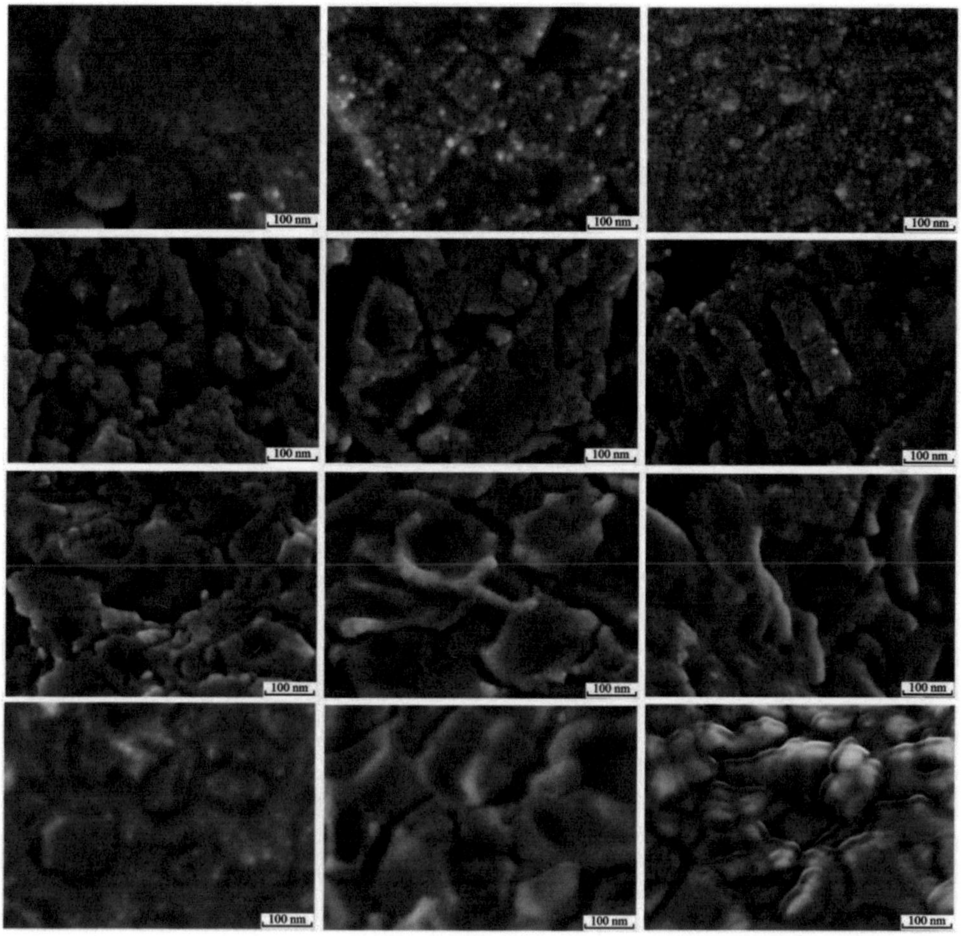

Figure 2.
The structural organization of palladium powders (T = 293 K), which transforms over time: separation into grains, their gradual compaction and the formation of flakes, their imposition on each other, and the formation of multilayer plates.

The system is an ultrafine mixture of metal particles and gas (air) filling the spaces between them. The morphostructural organization of metal nanoparticles formed during synthesis is represented by clusters of various configurations (up to 2 nm) and floccules (5–20 nm). The floccule structure is locally heterogeneous with a pronounced gradient of density decrease from the centre to the periphery. In the solid phase, the integration process continues. Floccules have a developed contact surface; smaller clusters, forming contact with them, significantly increase floccules in size, due to the expansion of a structurally loose boundary. A small amount of substance covers a large amount of space. Floccules with many interpenetrating and point contacts coagulate and form aggregates. The system begins the processes of self-organization and structuring. Floccule nuclei become denser, their flocculent boundary outgrowths and smaller free clusters grow together to form a quasi-amorphous cement (**Figure 3**). Later, the floccules are acquire a spherical shape. The substructure of the cementitious substance becomes indistinguishable.

The process of interaction of nanoparticles with an active unfolded surface, which provides multiple contacts, contributes to the development of their collective effects aimed at self-organization and structuring of the system. As a result,

Figure 3.
Evolutionary scheme of transformation of a solid phase system at the nanolevel: floccule coagulation → compaction with structuring elements → formation of structural elements (flakes).

new structural units form in the form of plates, flakes of various configurations from the total mass of loose flaky substance. Due to the compaction of the substance, their edges are often curled up, and the plates are separated (**Figure 2**). The substructure of the flakes in the form of stacking spheres at this stage remains clearly distinguishable. Observation of the dynamics of the system showed that at room temperature, the process of separation of particles of the second scale level is gradually activated, and larger fragments up to 200 nm are also individuated. They acquire clear boundaries as they condense. Their structure is composed of denser spheres cemented by a quasi-amorphous loose substance.

Developing over time, the processes of structural self-organization evolve to the consequent scale level, where the main structural units are no longer floccules and clusters but nanosized lamellar formations (~50–180 × 5–15 nm) which are formed from them. The process develops according to the planar contact deposition of nanoplates (flakes). This determines the morphostructural specificity of this stage of transformation. The scales continue to condense, the relief is gradually smoothed out and they become more closely adjacent to each other (**Figure 2**). As a result, a multilayer thin surface structure is formed of contact superimposed flakes, the substructure of which is still quite pronounced. Since the density of the substance at this stage of conversion is still quite low, the processes of compaction and structuring are decisive.

Thus, in the area of nanolevel scales, the solid phase evolution of palladium powders proceeds at three meso-levels of structural organization. The first meso-level is characterized by the integration of substance with the formation of clusters up to 2 nm in size (point, linear, volumetric of various configurations, the substructure is filamentous, flocculent) and floccules—segregations with a relatively dense core and indefinite flaky boundaries (5–20 nm). The second is the integration of floccules and smaller clusters, densification and structuring with the formation of nanoplates (flakes) (~50–180 × 5–15 nm). The third is the integration of flakes ("tiling" of flakes), the formation of thin multilayer submicroscopic plates with a movable easily transformable substructure. All the transformations described above took place in a thin layer of powder placed on a carbon tape substrate which is put on an observation table in an electron microscope for 6 months of observation.

A study of palladium powders in a sealed ampoule for 6 months at 293 ± 5 K showed that larger formations appeared in them in the form of branched micrograins, which are intergrowths of subgrains of rounded elongated shapes (**Figure 4**). The microstructure of subgrains is composed of nanosized polydisperse

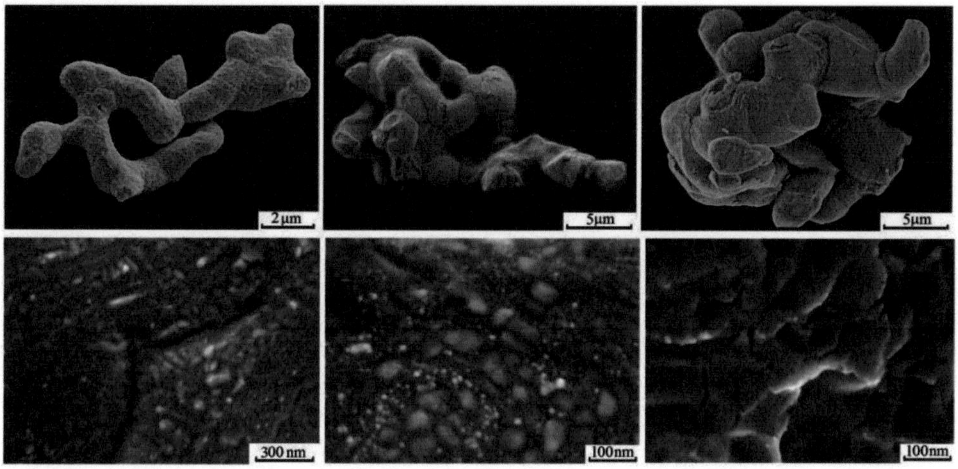

Figure 4.
Structural organization of palladium micrograins formed from powder after 6 months at 293 ± 5 K and their surfaces.

particles of varying degrees of compaction, on the surface of the subgrains without their mutual ordering and at the boundaries of intergrowth with other subgrains with polygonization elements (**Figure 4**). The particles are identical in their morphostructural features to the above-described particles which are formed in a thin layer of powder deposited on a conductive carbon tape.

4. Low-temperature solid phase self-organization processes

In finely dispersed palladium powders aged at negative temperatures, the integration processes take place at substantially higher rates than at a temperature of 293 ± 5 K. After 6 months of exposure to nanodispersed powders at a temperature of 258 ± 1 K, the occurrence of micrograins with sizes up to 250 μm is observed (**Figure 5**). The grains have a branched spongy structure, in the form of intergrowth of subgrains of rounded elongated shape with a diameter of 1–3 μm. The ratio of the volumes of voids and solids in the grains is comparable to each other. Many subgrains have end-to-end surface discontinuities, which indicate their hollow structure (**Figure 6**).

Studies of the structural organization of powder particles and palladium micrograins maintained at a temperature of 293 ± 5 K (T < Θ_D Pd) showed that the main processes here also include coagulation, densification, the formation of thin flakes and their aggregation with the creation of larger forms of thin multilayer plates, but in structural organization, they advanced significantly further than powders kept at a temperature of 293 ± 5 K (T < Θ_D Pd).

The surface of the plates (flakes) is more compacted, smoothed, with well-defined boundaries. Thin multilayer structural units gradually increase their area and, having reached micron sizes, begin to bend and twist, forming hollow, rounded, oval or tubular shapes (**Figures 6–8**). The multilayer structural organization of the plates ("tile laying" of the flakes) that form the walls of the hollow subgrains during bending provides them mobility due to the displacement of the layers relative to each other (**Figure 7**). With the increasing surface curvature, the upper plates move apart, revealing the inner layers with a quasi-amorphous surface. In this case, discontinuous discontinuities are often observed.

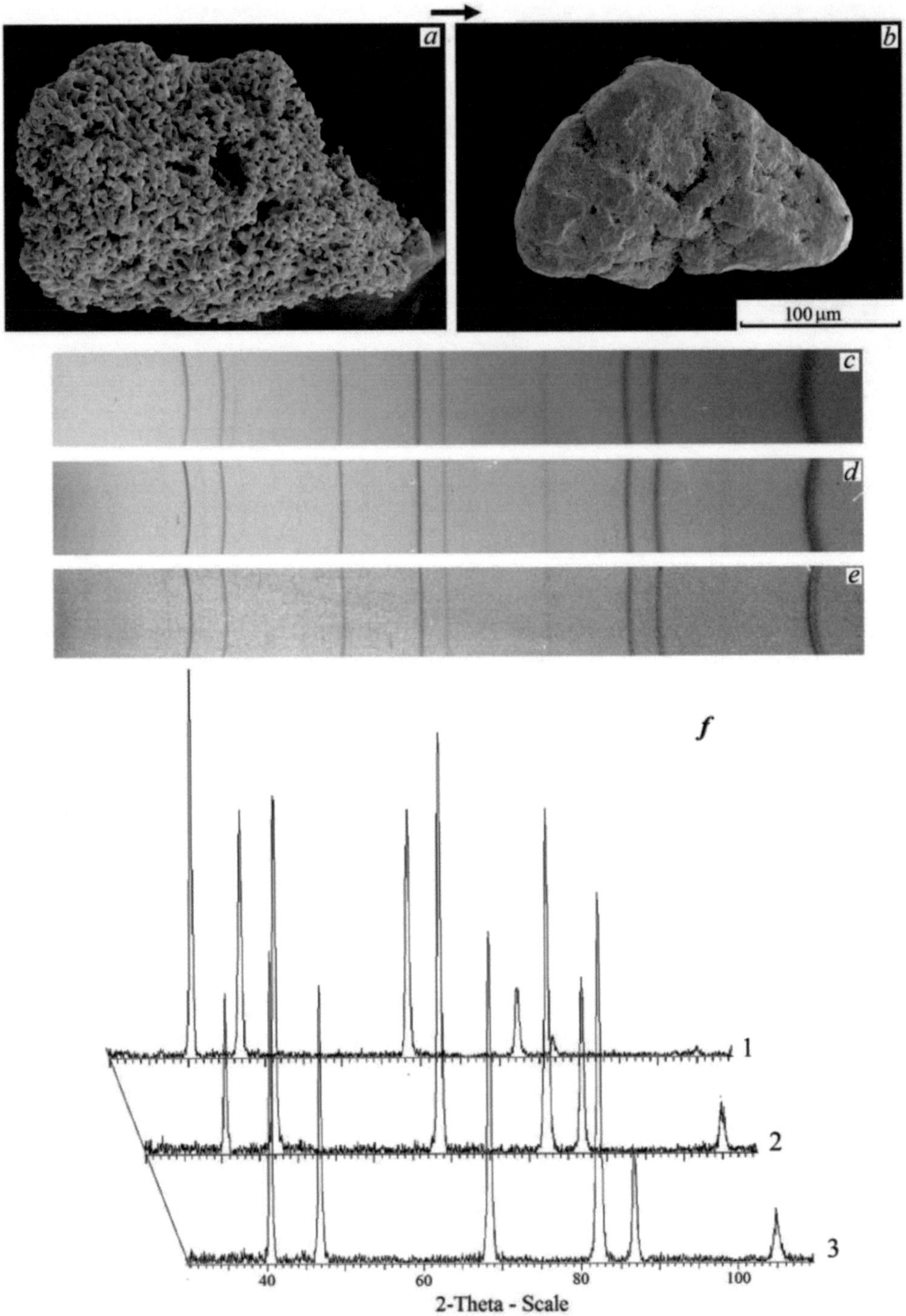

Figure 5.
Transformations of palladium micrograins at a temperature of 258 ± 1 K: sponge aggregate formed from nanopowder (after 6 months) (a), grain with a dense structureless surface and sponge internal structure (after 2 years) (b), X-ray diffraction patterns of palladium grain (after 6 (c), 12 (d), 24 months (e) at 258 K) and changes in the preferred orientations of crystallites in palladium grains at different stages of conversion (state after 6 (f1), 12 (f2), 24 (f3) months).

As a result of the processes of coagulation and compaction of the substance and the collective actions of nanoparticles of several scale levels, thin-walled hollow formations of rounded, oval, tubular shapes with a diameter of about 1–3 μm

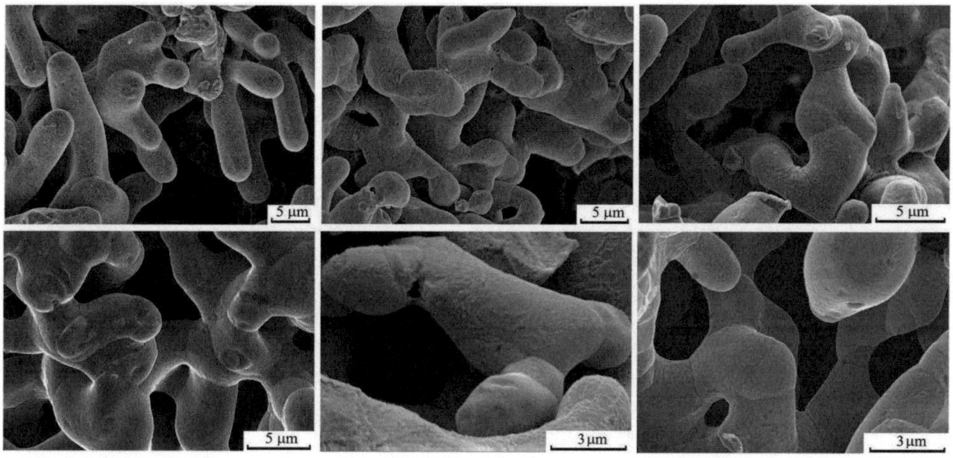

Figure 6.
Structural organization: sponge aggregate of palladium formed from nanopowder at a temperature of 258 ± 1 K (after 6 months).

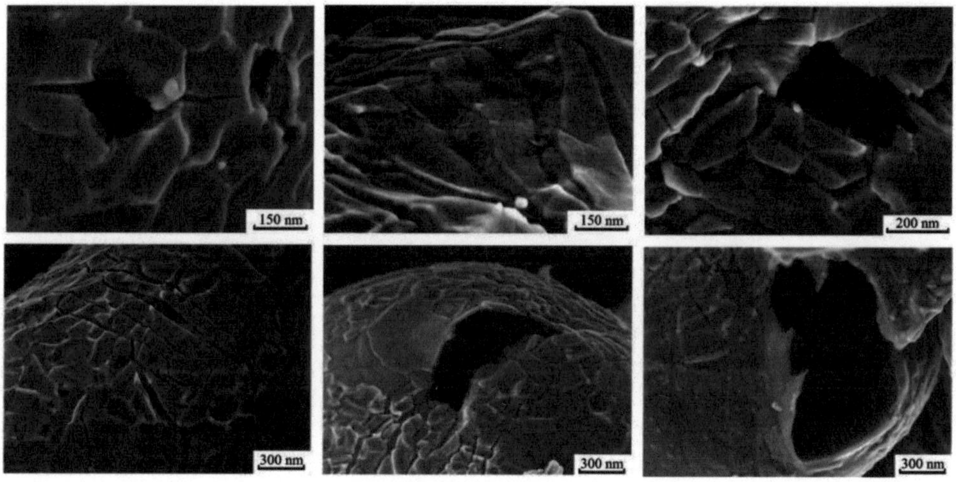

Figure 7.
Structural organization of multilayer planar intergrowth of plates (flakes). Destruction of the surface of multilayer lamellar formations during the formation of hollow forms.

are formed (**Figure 8**). These formations are structural units of the microscopic organization of matter. The structure formed in this way has an unfolded inner and outer surface. The diameter of the internal channels is close to 1 µm. The thickness of the multilayer wall is 30–60 nm.

By approaching each other, hollow micrograins grow together according either contact or bridge mechanism (**Figures 9** and **10**). The first mechanism is due to direct contact of curved surfaces (**Figure 9**). When the curved surfaces of two thin layers come into mutual contact, the layer thickness in the contact zone doubles. The scales composing the walls of the hollow grains come into motion due to the collective effects of the interaction. Mass transfer is directed away from the contact area (**Figure 9**). The outflow of particles causes opening of the cavity, the internal channels are combined and the wall thickness is levelled. The intergrowth process is not a sintering process with an increase in thickness in the contact area, but the unification and expansion of the thin channel internal structure.

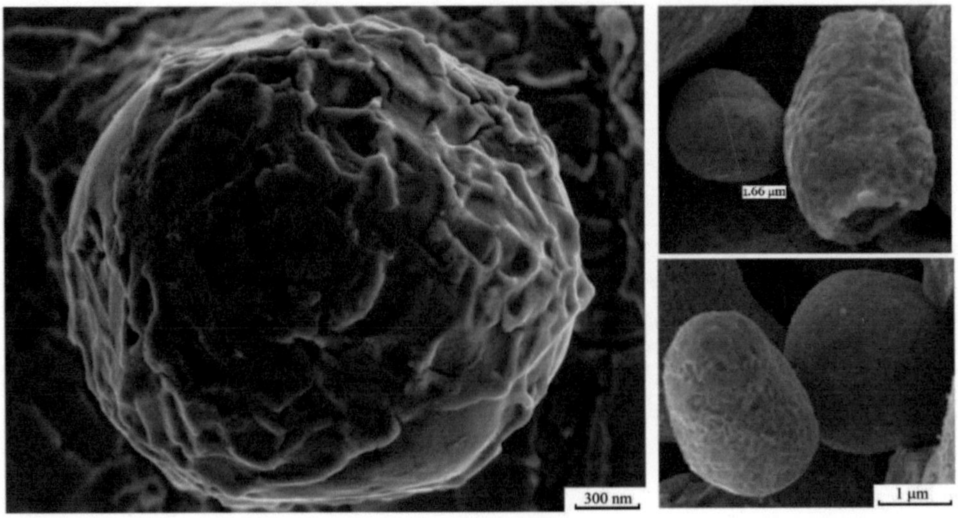

Figure 8.
Spherical and oval palladium micrograins formed as a result of aggregation of nanoparticles.

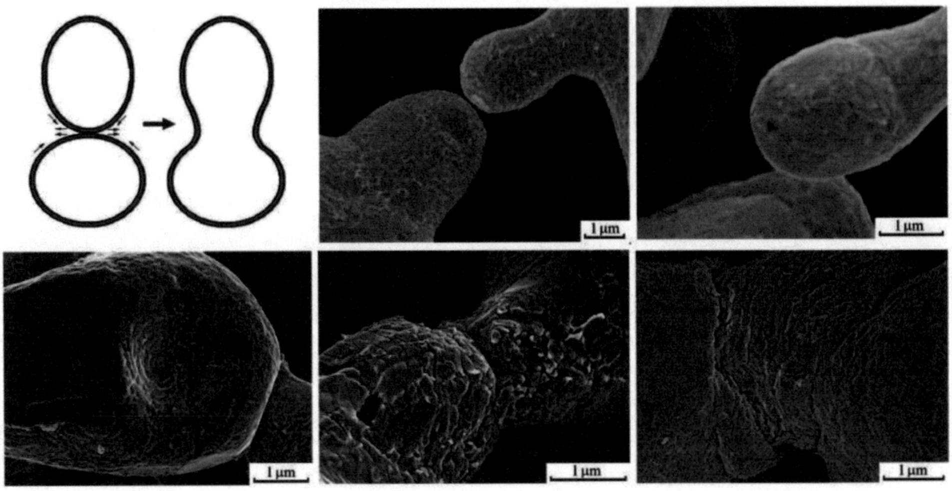

Figure 9.
Models of the contact mechanism of intergrowth of hollow masses. Contact interactions. Contact expansion and fusion of hollow structures.

The second aggregation mechanism, aimed at expanding the thin channel system, ensures the intergrowth of hollow thin-walled structures that do not contact each other. This mechanism is due to the effects of long-range bonds. When the curved surfaces of thin-walled (~30–60 nm) hollow formations are located at a distance comparable to the diameter of the internal channels (~1 μm) between the dispersed particles from which they are composed, collective effects aimed at combining the surface also appear. Collective actions of particles are manifested in mass transfer to the area of maximum approximation of surfaces (**Figure 10**). In this zone, a thin "bridge" begins to form, either in the form of a whisker-like outgrowth or a twisting effect forming an outgrowth cone that extends to contact with each other. After the formation of the bridge, mass transfer is activated. The contact bridge expands, and the transverse size becomes comparable with the size of the channels. The rupture structures and partings between the plates and scales are

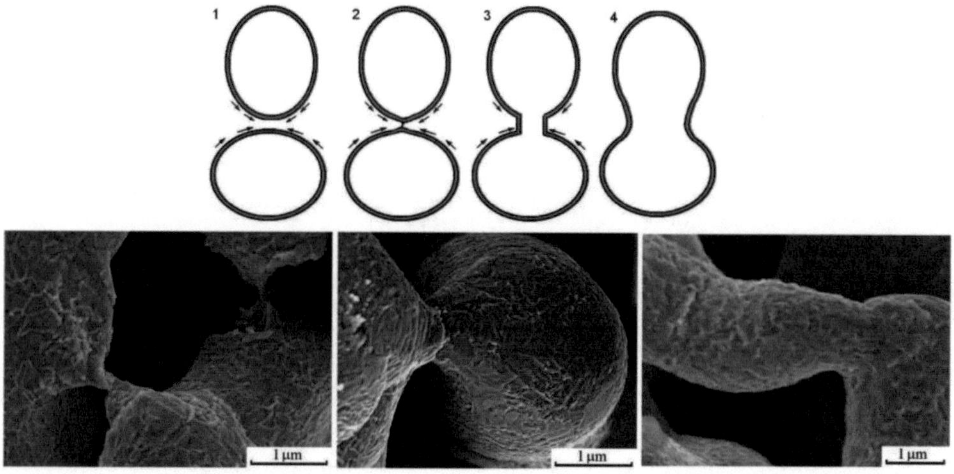

Figure 10.
The "bridge" mechanism of intergrowth of palladium micrograins, the model of intergrowth by the "bridge" mechanism. Formation of a contact "bridge". Expansion of the isthmus.

smoothed, forming concentric bands similar to growth zones. A single expanded channel system is being formed.

Thus, the palladium grains formed at this stage are a two-phase mixture consisting of a solid phase (Pd) and two types of voids (air). The first type is a void encapsulated in the internal cavities of micrograins and complex branched channels (~1 μm) formed when they merge. The second type is a spongy volumetric structure formed during the intergrowth of hollow microparticles. A void is almost evenly distributed over the grain volume.

The formed system has a developed inner and outer surface; it is energetically saturated and metastable. The structural configuration of the dissipative system involves its further transformation. The activation of interactions between hollow subgrains is characterized by two processes occurring simultaneously: the formation of new contacts and the destruction of previously formed ones. The constant movement of matter associated with the formation of new contacts often results in rupture of the surface and opening of channel structures. Viscous deformation discontinuities, shears and disruptions without visible deformations of the accretion boundaries are observed (**Figure 11**). Microcracks appear along the boundaries of intergrowth, new contacts form and open-channel cavities are closed. The system is transformed in a mobile way, reducing the volume of second type voids.

The processes of structural self-organization of particles in thin-walled bodies quickly form geometrically regular shapes (negative crystals) on the inner side of the channel wall, the outer surface is a multilayer packing of flakes and sometimes ordering effects are also observed in their arrangement, with the formation of face shapes (**Figure 12**). Particles whose structural organization is constructed in this way (tubes, spheres whose walls consist of nanostructured particles in a quasi-amorphous phase) are energetically saturated and have a high degree of freedom for further transformation.

Opening of internal channels causes collective effects in the system aimed at closing them. The process is due to transformations of the substance in the area of the open hole and depends on the mechanism of rupture. If a rupture occurred due to tensile forces, leading to the formation of a tension zone and a thin isthmus at the time of the rupture, the boundaries of the rupture have thin ragged edges, with a significant number of nanosized separation protrusions. Such discontinuities are

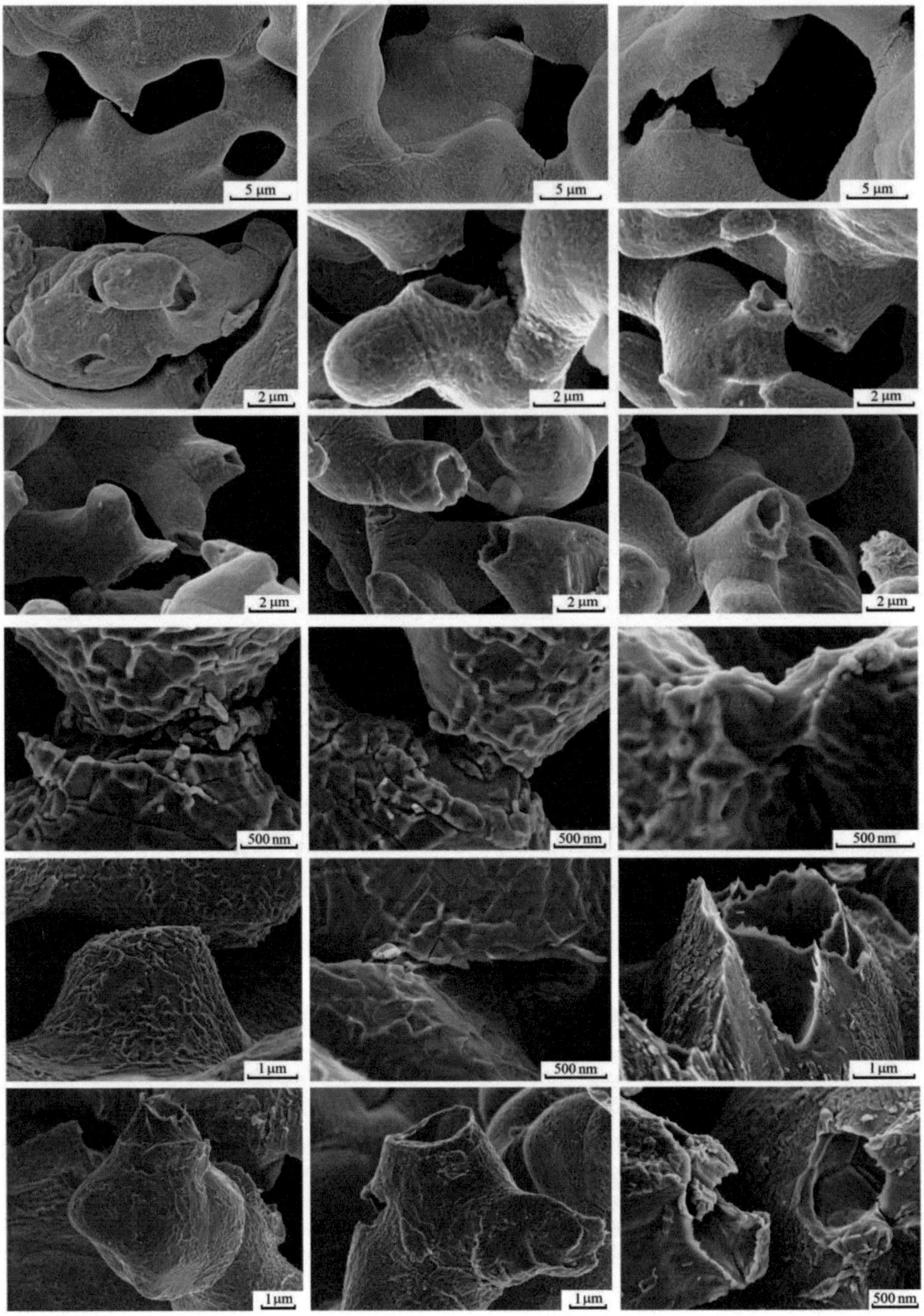

Figure 11.
Various types of strain breaks in palladium grains: after 6 and 12 months at 258 ± 1 K.

closed as a result of the convergence of the walls and smoothing of the displacement formed during the rupture (**Figure 13**). When the size of the holes formed, for example, as a result of shear forces, corresponds to the diameter of the channel, their closure occurs mainly due to the effects of quasi-viscous flow (**Figure 13**). Incrusted structures grow, gradually closing the hole. This type of closure is most common in the studied system. The convergence of the walls in the rupture

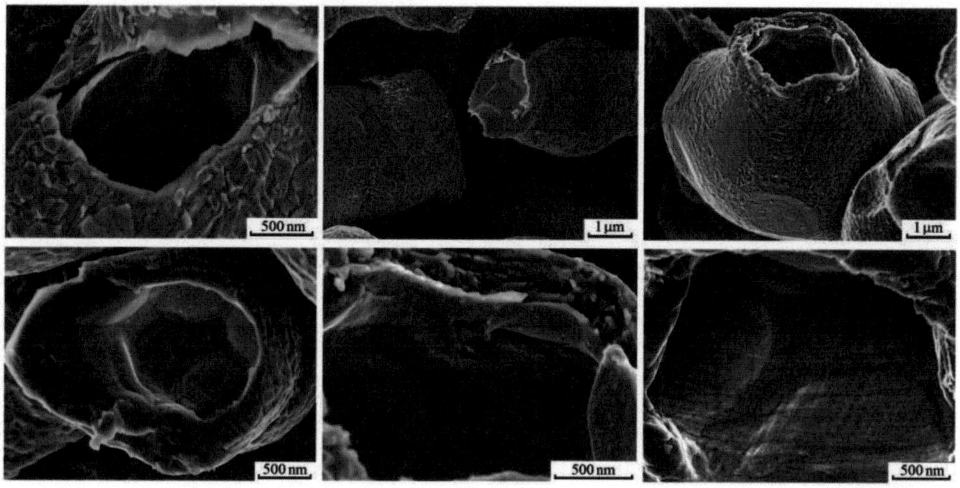

Figure 12.
Recrystallization of the inner wall of hollow channel structures in palladium grains and elements of ordering of the outer layer.

boundary area is also observed; the outer diameter of the hole decreases until contact interactions between the particles of the surface layer are established, causing the processes of restoring the integrity of the outer layer and the formation of a common shell (**Figure 13**).

The development of the compaction process over time leads to the convergence of the boundaries of hollow subgrains. When there are several surfaces of hollow formations at a critical distance at a time, a unified surface is formed forming structural units of the next scale level in the form of intergrowth of rounded bodies with sizes of 3–30 μm, which gradually form curved surfaces preserving the boundaries (**Figure 14**). Their surface also has a scaly lamellar structure with elements of polygonization, and there are residual pores of the second type, which gradually decrease in size.

The main mobile structural units of the surface layer of subgrains are scaly lamellar formations. They do not lose their structural identity for a long time (**Figure 15**). After the merging of the hollow subgrains and the formation of rather extensive zones with denser intergrowth, the process of surface transformation is activated. Initially, the borders and shapes of the scales are structurally separated, and then spicule-like elements appear in the structure of the scales with the ordering effect in their location; the relief contours are softened and smoothed. Then, scales are flattened and boundaries are merged. Consequently, zones with structural corrugated bands are formed (**Figure 16**) which are gradually smoothed out levelling the surface. On the surface, the elements of polygonization are manifested, with the development of the process at the neighbouring subgrains.

In the processes of surface transformation, the key role is played by the size effect. The collective actions of ultrafine particles determine the order of transformations that develop over time. In the initial stages, in spite of the volumetric filling of space with particles, ordering has a planar distribution. Consequently, the transformation process goes to the sub-grain level, and then it is activated in the contact zones. With the coalescence of micrograins, the intergrain zones are also involved in the collective processes of surface transformation, the boundaries gradually disappear, and the structural organization of the surface becomes unified (**Figure 16**). At later stages, the surface topography is smoothed and acquires soft forms, and then banded and other forms with a relatively smooth relief appear. A structure identical

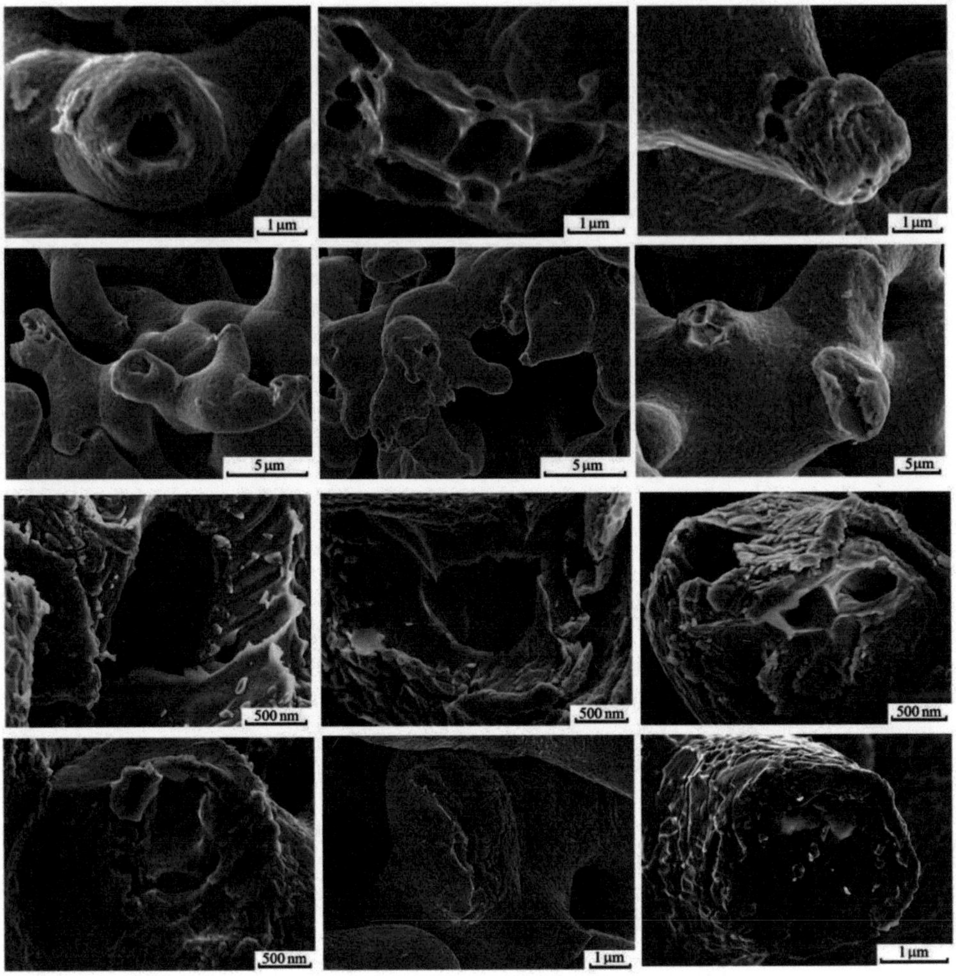

Figure 13.
Closures of discontinuities in channel structures in palladium grains: walls convergence and displacement smoothing, quasi-viscous flow effect and recrystallization.

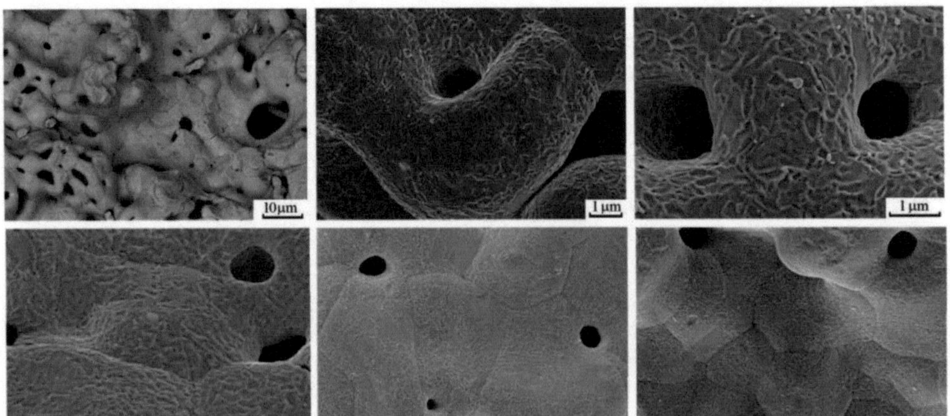

Figure 14.
Different densities of micrograin intergrowths and stages of surface transformations.

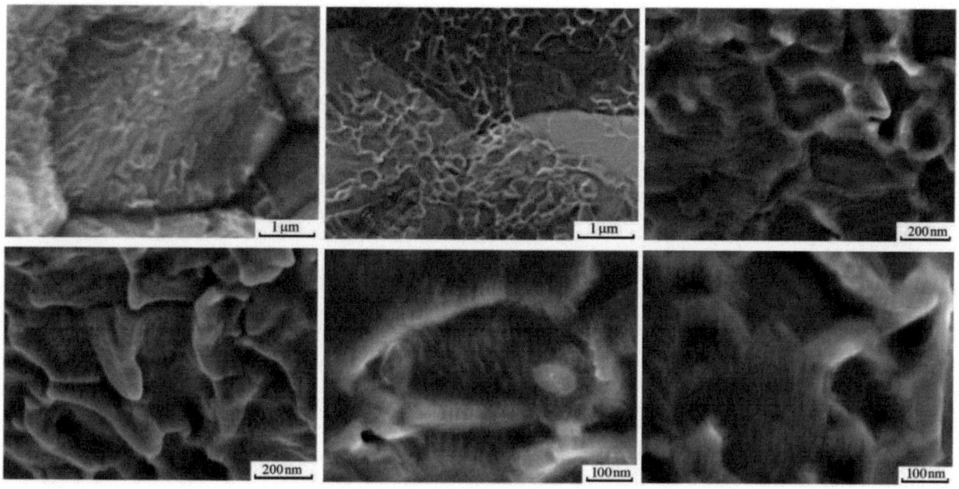

Figure 15.
Transformation of the surface shell of hollow grains, quasi-viscous flow effect and recrystallization.

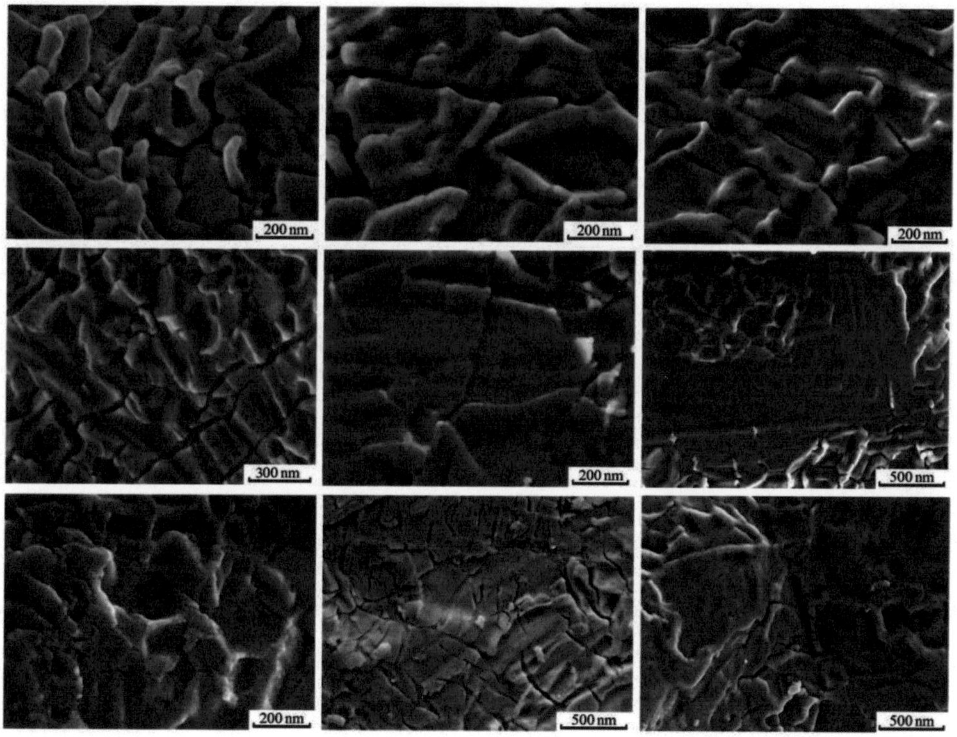

Figure 16.
Extension of the area of smoothing of the surface and with the ordering of structural elements.

to the structures of the growth planes is revealed (**Figure 17**). A relatively stable state is organized in which slow processes occur, aimed at ordering between the particles of a thin layer—the shell. Subgrains lose their structural identity; former contact zones acquire smoothed shapes.

The next important stage in the structural transformation of palladium micrograins is associated with the creation of a unified grain shell as a whole.

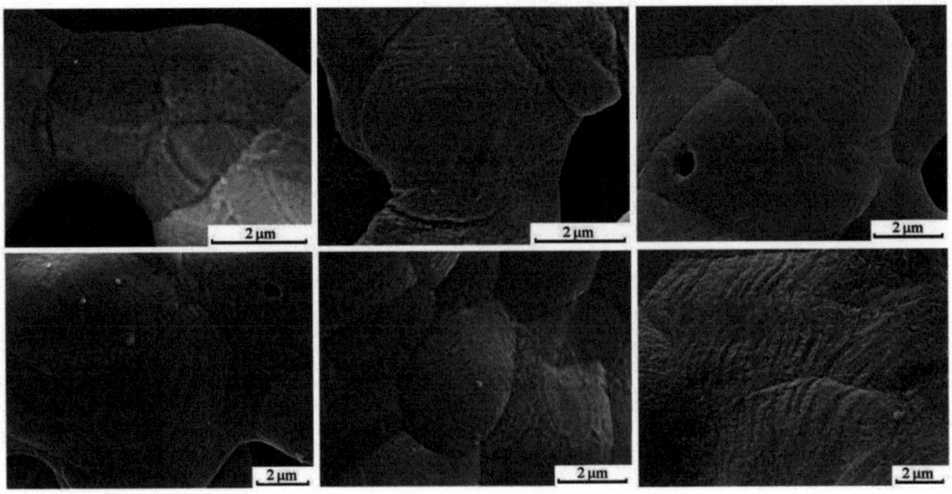

Figure 17.
Ordering within individual grains with the occurrence of growth planes and twins.

After the formation of a micrograin (50–300 µm) of a spongy structure, the compaction processes contribute to its transformation as a whole, separating it from the rest of the fine powder. Since the aggregation and occurrence of grains have a progressive character with the simultaneous presence of both a finely dispersed phase and already formed grains in the system, particles of various sizes and degrees of compaction were found at various stages of the study. Transformation processes are aimed at the individualization of each individual grain, the creation of a common shell (the microencapsulation effect). When the displacement of voids of the second level is activated, resulting in compaction of the substance, the most active merging of particles occurs on the grain surface. A grain, as a whole object, tends to acquire a single, generalized surface. The individuality of the substructural components is first smoothed out due to effects similar to a viscous-flowing state, and then sod-deformed, smoothed, worn-out forms are formed, gradually closing the internal porous channel structure (**Figure 18**). The organized outer shell loses smoothness, becomes fine-grained and is worn smooth. The microencapsulation effect is aimed at organizing a common surface on which one can only detect relics of a structural organization that has undergone significant evolutionary transformations. The surface acquires a structure that is described as rounded with deformation elements (**Figure 18**). However, despite the presence of an external dense shell, the internal organization of the grain remains nanoporous.

Stabilization of the temperature regime at temperatures below the Debye palladium characteristic temperature favours the occurrence of preferred orientations of crystallites in palladium grains (**Figures 1**, **5** and **19**). If the spongy grains, whose crystallites do not exceed 1 µm, are kept at room temperature, no increase in crystallite size will be recorded. Radiographs are characterized by sharp lines (**Figure 19a**), and the intensity distribution is identical to the theoretical radiograph. The X-ray structural characteristics of palladium grains maintained at various temperature conditions are presented in **Table 1**.

The unit cell parameters for palladium grains transforming at room temperatures are approximately 3.882 Å. This value is reduced relative to the unit cell parameters of native palladium, which indicates the incompleteness of the transformations in the system of its metastability. And for samples aged under freezing temperatures, the parameter values are as close as possible to the values for native palladium. For negative temperatures close to the Debye characteristic

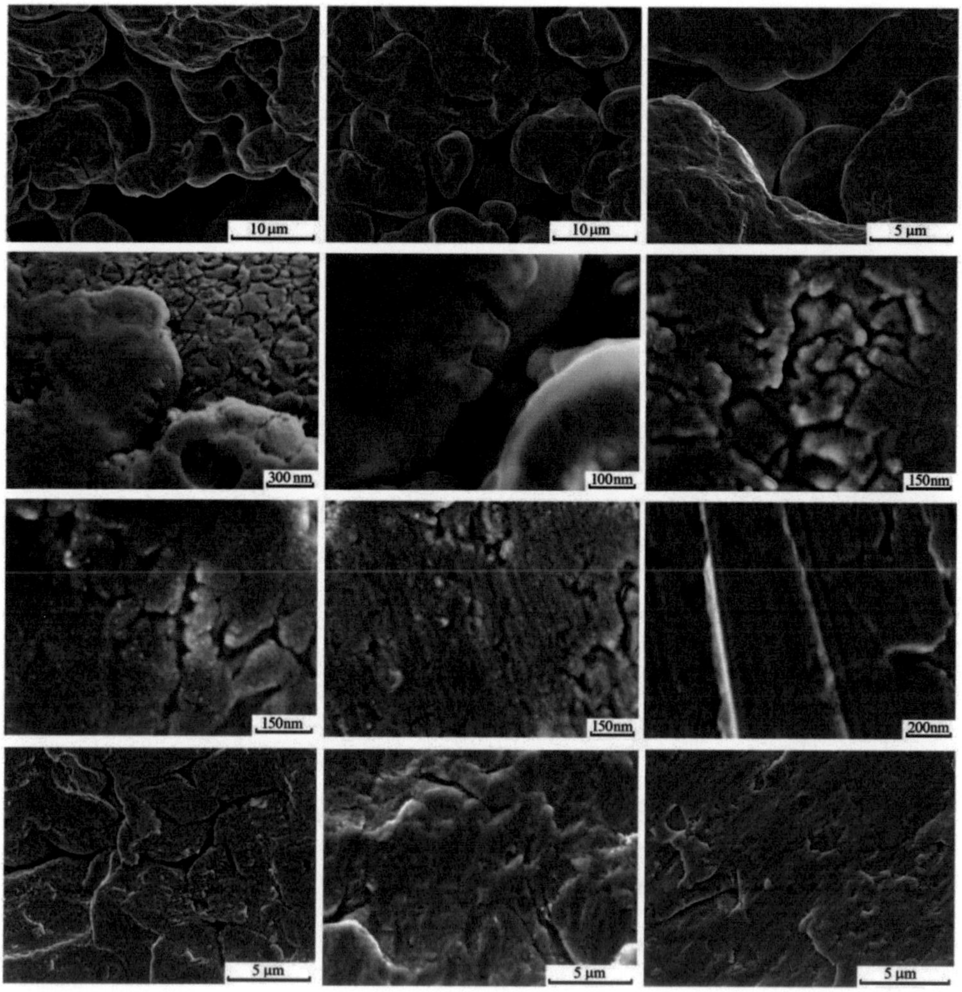

Figure 18.
Grain encapsulation and surface transformation.

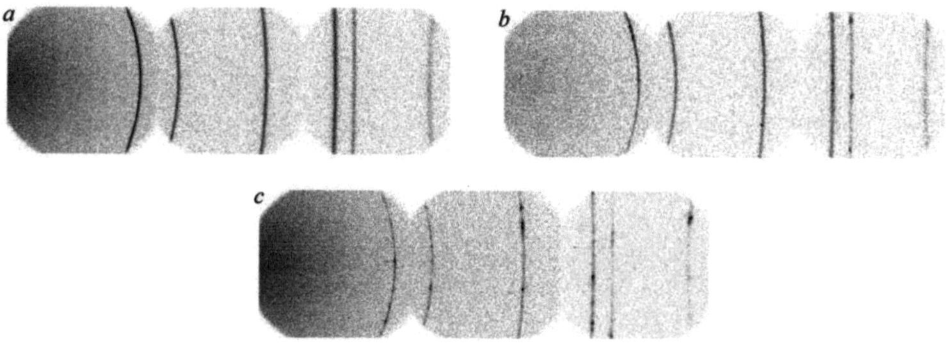

Figure 19.
Radiographs characterizing the changes in the preferred orientations of crystallites in palladium grains, aged at 293 ± 5 K (a), at 258 ± 5 K (b) and at 77 K (c) (shooting at the Collective Use Centre of the Far Eastern Geological Institute of the Far Eastern Branch of the Russian Academy of Sciences).

temperature of palladium, the ordering process is accelerated, and the crystallite sizes increase. It is likely that further transformation processes will lead to a single crystal state. However, the presence of internal channels/voids leaves the system

Sample		For all available reflections				In the precision area		
		a ± 0.001 Å	V, Å^3	Δd*	Mn**	a ± 0.0001 Å	Δd*	Mn**
Pd 00-046-1043		3.89019(2)				3.89019(2)		
Pd 293 ± 5 K	1	3.882	58.482	0.00276	80.044	3.8831	0.00027	200.688
	2	3.882	58.512	0.00351	68.080	3.8839	0.00019	292.893
	3	3.881	58.467	0.00244	85.800	3.8825	0.00040	164.194
Pd 258 ± 1 K	1	3.887	58.717	0.00135	85.196	3.8871	0.00087	62.679
	2	3.890	58.870	0.00163	98.430	3.8911	0.00042	135.758
	3	3.890	58.857	0.00112	165.352	3.8906	0.00022	274.520
Pd 77 K	1	3.890	58.870	0.00096	153.662	3.8909	0.00007	906.799
	2	3.889	58.816	0.00072	183.220	3.8894	0.00024	233.847
	3	3.888	58.765	0.00125	127.889	3.8885	0.00034	162.065

Δd, average deviation of the calculated interplanar distances off the measured ones;*
*Mn**, reliability criteria after de Wolf.*

Table 1.
X-ray structural characteristics of palladium grains aged in various temperature conditions.

open. A significant decrease in temperature (77 K) promotes rapid encapsulation (**Figure 20**), the formation of a dense shell, in which ordering occurs only at the micro level. That is, the processes of separation of each individual grain in this state are predominant over the processes of general ordering.

Thus, studies of interactions between ultrafine particles of palladium in free contact showed that their aggregation leads to the formation of thin channel structures with a channel diameter of ~1 μm and a wall thickness of ~30–100 nm. The sizes of the coherent dissipation blocks and the wall thickness of the channel structures are identical; their inner surface has flat faces in the form of negative crystals, which indicates the ordering between the particles composing them. The established effect of enlargement of coherent dissipation blocks and the occurrence of preferential orientations are probably associated with ordering in the arrangement of channel structures and their alignment in a certain order. Microencapsulation favours an orderly distribution of voids in the grain volume.

Dissipative processes in dispersed objects contribute to the development of long-range coherence in the system [13]. The determining parameter in this process, in our opinion, is the ratio of the sample holding temperature to the Debye characteristic temperature of palladium Θ_D = 274 K. The Debye characteristic temperature is the temperature at which crystal lattice vibrations of all possible frequencies are realized. A further increase in temperature does not lead to the occurrence of new vibration modes but only leads to an increase in the amplitudes of existing ones, that is, the average vibration energy increases with the increase in temperature. At temperatures of the system $T < \Theta_D$, the atoms oscillate near a certain equilibrium position, the intensity and amplitude of which depend on the difference between the Debye characteristic temperature of the substance and the temperature of the system ($\Delta T = \Theta_D - T$). According to Debye, with a decrease in temperature in the area $T < \Theta_D$, the energy of each individual vibrator decreases; moreover, the vibrations of more and more new vibrators gradually decrease.

Thus, under conditions when $T < \Theta_D$, interatomic interactions become predominant. Long-range bonds are established between the atoms of matter; the collective actions of atoms are aimed at ordering and tending of the system towards the perfection of its crystal structure. The structure of the substance becomes denser.

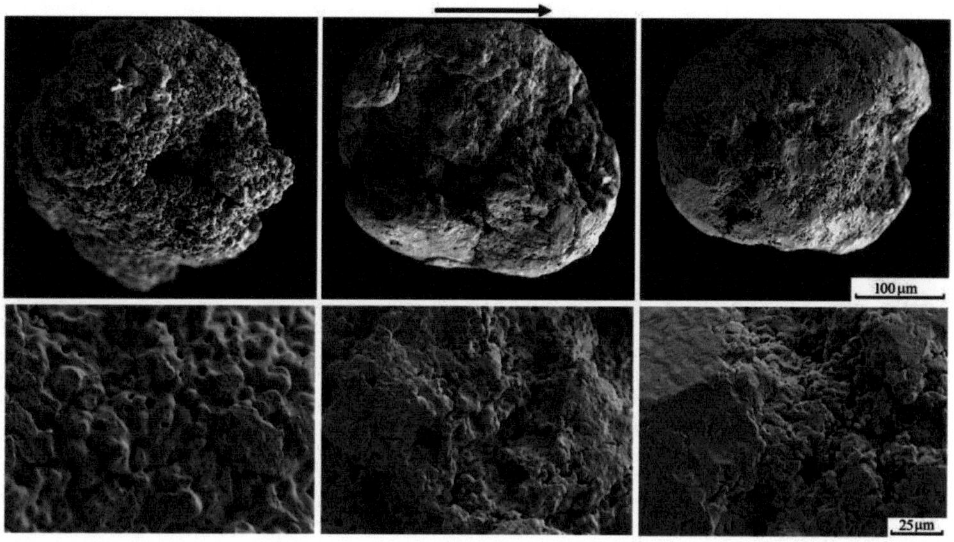

Figure 20.
Cryotemperature transformations of palladium grains with structure compaction at 77 K: 6, 12 and 24 months.

Further lowering the temperature reduces the amplitude and energy of the natural vibrations of the lattice, thereby contributing to even greater structuring of the substance.

5. Conclusion

Thus, studies of interactions between finely dispersed palladium powders that are in free contact under dissipative conditions of low temperatures indicate the group behaviour of particles at various scale levels. Structural aggregation takes place in stages, with the allocation of an aggregation field area at each scale level:

- Lack of structure
    - Small-scale clustering
- Prestructuring
    - Medium-scale aggregation and divergence of cluster types in the aggregate
    - Large-scale aggregation
- Structuring
    - Aggregate formation and shell formation (microencapsulation)
    - Recrystallization at the subgrain level
    - Recrystallization at grain level

In the collective effects of self-organization, periodic activation of mass transfer is observed, in which nanoparticles of various mesoscale structure organization are involved. The solid phase finely dispersed substance evolves mobile in the process of dissipation of surface energy, which takes on new forms at every scale level of the organization while maintaining the structural identity of its constituent particles. The manifestation of a multiscale structure, close relationship and mutual influence of various scale levels is one of the key points in the processes of mass transfer in contact interactions.

Conflict of interests

The author states having no conflict of interests.

Author details

Veronika Ivanovna Rozhdestvina
Institute of Geology and Nature Management FEB RAS, Blagoveshchensk, Russia

*Address all correspondence to: veronikapts@yandex.ru

References

[1] Bönnemann H, Richards RM. Nanoscopic metal particles—Synthetic methods and potential applications. European Journal of Inorganic Chemistry. 2001;**10**:2455-2480. DOI: 10.1002/1099-0682(200109) 2001:10<2455::AID-EJIC2455> 3.0.CO;2-Z

[2] Saldan I, Semenyuk Y, Marchuk I, Reshetnyak O. Chemical synthesis and application of palladium nanoparticles. Journal of Materials Science. 2015;**50**(6):2337-2354. DOI: 10.1007/ s10853-014-8802-2

[3] Cookson J. The preparation of palladium nanoparticles controlled particle sizes are key to producing more effective and efficient materials. Platinum Metals Review. 2012;**56**(2):83-98

[4] Redon R, Rendon-Lara SK, Fernandez-Osorio AL, Ugalde-Saldivar VM. Aerobic synthesis of palladium nanoparticles. Reviews on Advanced Materials Science. 2011;**27**:31-42

[5] Bradley JS. The chemistry of transitional metal colloids. In: Schmid G, editor. Clusters and Colloids: From Theory to Application. VCH: Weinheim; 1994. pp. 469-473

[6] Xiong Y, Xia Y. Shape-controlled synthesis of metal nanostructures: The case of palladium. Advanced Materials. 2007;**19**:3385-3391. DOI: 10.1002/ chem.20040092

[7] Oduro YH, Tam W, Tsang E. Chemical methods for preparation of nanoparticles in solution. Handbook of Metal Physics. 2008;**5**:385

[8] Dahman Y. Nanoparticles (Chapter 5). Nanotechnology and Functional Materials for Engineers A Volume in Micro and Nano Technologies. Elsevier Inc. 2017:93-119

[9] David R, Marchal P, Klein J-P, Villermaux J. Crystallization and precipitation engineering—III. A discrete formulation of the agglomeration rate of crystals in a crystallization process. Chemical Engineering Science. 1991;**46**:205-213

[10] Hotze EM, Phenrat T, Lowry GV. Nanoparticle aggregation: Challenges to understanding transport and reactivity in the environment. Journal of Environmental Quality. 2010;**39**(6):1909-1924. DOI: 10.2134/ jeq2009.0462

[11] Geguzin JE. Physics of Sintering. Moscow: Nauka; 1984. pp. 312

[12] Blendell JE. Solid-state Sintering. In: Encyclopedia of Materials: Science and Technology. Elsevier Ltd. 2001. pp. 8745-8750

[13] Kondepudi D, Prigogine I. Modern Thermodynamics: From Heat Engines to Dissipative Structures. Chichester, UK: Wiley; 1998. pp. 486